Scrooge's Cryptic Carol

Scrooge's Cryptic Carol

Visions of Energy, Time and
Quantum Nature

Robert Gilmore

COPERNICUS
AN IMPRINT OF SPRINGER-VERLAG

© 1996 by Sigma Press.

Published in the United States by Copernicus, an imprint of
Springer-Verlag New York, Inc.

Copernicus
Springer-Verlag New York, Inc.
175 Fifth Avenue
New York, NY 10010
USA

Library of Congress Cataloging-in-Publication Data
Gilmore, Robert, 1941–
Scrooge's cryptic carol : visions of energy, time and quantum
nature / by Robert Gilmore.
p. cm.
Includes bibliographic references and index.
ISBN 0-387-94800-7 (hardcover : alk. Paper)
1. Scrooge, Ebenezer (Fictitious character)–Fiction.
2. Science-History–Fiction.
3. Quantum theory–Fiction
I. Title
PS3557.I4595S37 1996
813'.54–dc20 96-15529

Manufactured in the United States of America.
Designed by A Good Thing.

9 8 7 6 5 4 3 2 1

ISBN 0-387-94800-7 SPIN 10539378

To my friends
past and present,
in the Bristol Physics Department

Contents

Introduction

Where Has All the Mystery Gone?

Scrooge minded his own business, which is fair enough. If you do not mind your business, who will? It does not follow, however, that he had to be insensitive to the mystery of the world about him.

The Universe *is* a mystery. Probably it always will be to some degree, but we can still attempt to understand it. One of the most marvelous things about the universe is that much of it does seem to be intelligible, particularly through the language of mathematics.

The complaint is sometimes made that Science (with a capital S of course) destroys the poetry of life by removing all the wonder and mystery from the world. The people who say this are in general not scientists. I do not believe that an understanding of any aspect of the physical world can remove the wonder. It may remove some mystery, but in doing so it will introduce new mysteries. Science can help us to understand some aspects of the world, but in the process you find that you do not understand things of which you were not even *aware*. It may seem unsatisfactory that the task of understanding seems doomed to be never-ending, but though we may not know everything about the physical world, it still behooves us to know as much as we can. We are stuck in the world after all, and simple nosiness (or wonder, if you prefer) would require that we examine it.

In the following pages, Scrooge is conducted by ghostly spirits on a survey of the physical world, which occurs in three separate visitations. The first is, in a sense, that of "Science Past," a vision of energy and entropy, quantities well known to the science of the nineteenth century or before. This vision speaks of the science of "being" rather than "becoming." It speaks of energy conservation and thermodynamic equilibrium, of Nature's progress toward a changeless state.

The second vision deals broadly with time. It cannot accurately be described as a vision of Science in the present moment, and indeed one of the messages is that the present moment is a personal thing and is not universal. This discourse deals with motion, with change and with nonequilibrium thermodynamics, which can result in the creation of order. The equilibrium thermodynamics of the Victorian era is not the last word on the subject. In a *sense* perhaps it is the last word because it predicts the Heat Death of the universe: the dreary end of everything in a state where nothing significant will happen. You might then say that nonequilibrium thermodynamics is the *first* word, in the Biblical sense of creation.

The third vision deals with the quantum aspect of nature. This is not really Science Future since much of the development took place around 1930; but it is definitely Science Weird. Quantum physics has now become a reliable workaday tool for physical calculations, but if you want mystery in the universe you need look no further.

The book is written in a narrative form, in which the Spirits conduct Scrooge around the universe and show him visions to illustrate the theories of physics. The message of the Spirits is not always consistent, as they are describing the picture of the world given by science, our science. The picture of the world given by science changes slightly from time to time, and so some of the statements of the earlier visitors are modified by those who follow. That is in the nature of Science. Earlier theories, even those well supported, are found in certain circumstances to be only approximations to newer (and usually stranger) pictures of the world. At the end of this sequence of visitations you might be tempted to ask "What is the final truth?" It is a fair question, but how should I know the answer?

In the course of Scrooge's travels with the various Spirits he frequently moves great distances through time and space or shrinks to a minute size. None of this is actually possible. The demonstrations of the Spirits are possible only in the imagination but the message they convey is, as far as I know, true.

Scattered throughout the chapters are short sections of text enclosed in boxes. These outline, in a more direct and prosaic form, specific results which relate to the chapter.

Mathematics

This book does not include any mathematics in the body of the text, but a little is included in some of the information boxes mentioned above. It is often claimed that any mathematics in a book intended for a nonspecialist readership is certain disaster. This seems a pity, since scientists do not use mathematics just so that they may confuse the uninitiated. Mathematical notation gathers up the quantities concerned in a compact notation and keeps them together in front of the reader's eye.

The equation $\mathbf{p} = mv$ shows at a glance that mass m and velocity v are equally involved in determining the size of an object's momentum $\mathbf{p}$. The equation $E = \frac{1}{2}mv^2$ for kinetic energy shows that here again m and v are involved, but now the velocity v has the dominant role. Writing v^2 instead of $v \times v$ emphasizes the fact that *the same thing* is being multiplied by itself.

That is about as complicated as mathematics will get in this book. You could put a great deal more precision than this into a mathematical expression. For example, an expression like

$$\frac{\partial}{\partial t} \iiint_{\Delta_t} P(r,t)d^3r = \iint_{S_\Delta} J_P \cdot \hat{n}\, dS$$

would convey a lot of information to those who are used to the notation, but it is not for us.

Above all, the mere presence of mathematics is not contagious. You do not have to read the equations if you do not want to, and the explanations in the text will be the same as they would have been if the mathematics had not been added. So that you may, if you wish, creep past on tiptoe without rousing the mathematics that is present, each instance is labeled with a warning sign,

This shows that a mathematical equation follows.

❦
Prologue

Marley's Ghost

Marley was dead: to begin with. There was no doubt whatever about that. Kevin Marley, upwardly mobile young financial adviser, had taken a corner too quickly in his new Porsche and abruptly found himself to be even more mobile than he had expected. Whether this mobility was upward or downward, however, no one was in a position to say.

Marley was certainly dead. He was dead as a doornail, though I do not know of my own knowledge what is so particularly dead about a doornail. Perhaps in his case one might say that he was as dead as a policy with a missed premium. His partner, Scrooge, knew that he was dead. He had been his sole executor, his sole legatee, his sole friend, and his sole mourner. Even so, he had not been so dreadfully cut up by the sad event, since he was now the sole partner in the firm of "Marley and Scrooge."

One evening Scrooge sat working late in his office. He had had a most profitable day, as indeed he took care that most of his days

1

were; and he was absorbed in recording the accounts of the day's transactions. He sat behind his beautiful and efficient desk like a statue carved from ice, impeccably dressed in his handmade business suit and coldly reviewing the day's profit and loss on a computer terminal. As was usually the case, he found much profit and little loss, though he noted each with the same cold attention. People often said that there was much of the computer about Scrooge in that they both pursued their ends with the same cold, calculating efficiency; but computers have been known to play games, and Scrooge never did. His secretary had left the outer office some time before and so he had no intimation that he had a visitor until he was suddenly accosted.

"Good evening cousin!" cried a cheerful voice. "Still at work making money, I see. Would you care to contribute some of it to a worthy cause?" It was the voice of Scrooge's cousin, a teacher of science from a local school. His appearance was in considerable contrast to that of Scrooge himself. While Scrooge was dressed in designer shirt and tailor-made suit, his cousin wore baggy flannels and a tweed jacket with patches sewn on the elbows. Scrooge felt that he had made little enough of his life to be so confoundedly cheerful.

"Would you care to contribute to the Save British Science fund?" continued this unsatisfactory cousin. "It is a movement to promote interest in science and to support scientific research and education, so that we shall not become a backwater nation in which there is little understanding of scientific principles. We do not want the great discoveries of the past, which have been made with so much ingenuity and effort, to be lost to the people of tomorrow. Come now, cousin, even a small contribution would help, and I know that you can afford it. You're rich enough."

"It's not my business," Scrooge returned. "It's enough for a man to understand his own business and not to interfere with other people's. Mine occupies me constantly. If you want money to support your precious science you should go out into the marketplace and earn it. You would be better if you concentrated on your own career. You're poor enough. People like you have lived for too long in your ivory towers and you expect others to support you. You will have to learn to live in the real world!"

After a few further exchanges along these lines his cousin departed much disheartened. Scrooge finished his work with an improved opinion of himself, as the darkness thickened further outside. Shortly thereafter he left, carefully locking up the office behind him but leaving the fax and telephone answering machines unceasingly vigilant on his behalf for any word of further business. He walked down the street from the door of his building and descended as usual into the nearby station of the London Underground.

As on so many other evenings, he waited on the platform until his train came rushing out of the tunnel at one end. As it passed, he glimpsed the destination board at the front of the train, and in the instant that it passed he could have sworn that its stated final destination was not HEATHROW, as he might have expected, but instead the destination board read HEAT DEATH. He was momentarily startled, since he was not a man given to such fancies. Perhaps he had been working a little too hard. Maybe he should treat himself to a break; somewhere sunny in five-star luxury.

The train journey passed without incident, and soon he emerged from the station closest to his fashionable apartment, onto a street brightly lit by a progression of streetlights below the dark nighttime sky. The way to his home took him past the windows of a large store, now closed for the night but still brightly illuminated with all its wares on show. Most prominent of these was a bank of television screens, each one showing the same program. Piled one on top of the other and ranked in rows the full width of the window, each one showed an identical picture of a talking head although no sound was audible through the window. Again and again was repeated the face of some lesser-known politician, his inaudible words little loss to the passerby. As his eye passed idly over this expanse of animated wallpaper, Scrooge realized with a shock that one face was different. The face in the screen nearest to him was that of his dead partner Kevin Marley.

He was looking directly at Scrooge and appeared to be speaking most earnestly but, strain as Scrooge might, no sound of this message came through the window. Now quite convinced that he was urgently in need of a vacation, Scrooge hurried on, uneasily aware of Marley's face, which was switching from screen to screen to keep alongside him as he strode past with quickening step.

He stepped along smartly, anxious to be home and in the consoling warmth and comfort of his own apartment. Soon he turned the corner into the short street that led to his door. The street was narrow but well lit by a row of lamps which stretched all the way down to the far end. As he passed below the first of these it abruptly went out and left him walking through a pool of shadow, although the way in front was still brightly lit. He marched determinedly on and was just composing the letter of complaint that he would write to the local authority as he passed under the next light. This promptly extinguished also, and he noticed that, at the same time, a light at the far end of the street had darkened too. He hurried on, a little more nervously, and as he passed below each light it went out, every extinction being matched by its twin at the other end of the street. As he moved he was thus effectively trapped in a narrowing pool of light, which, he came to realize, was centered on his own doorway.

Soon he reached his door, just as the final lights went out, and left him in a darkness such as he had not encountered for a long time. He managed to get his key into the lock but just before opening the door he chanced to look up. Above him he saw the night sky, no longer masked from him by the blaze of artificial light, but

deep and clear as eternity. Perhaps this was the greatest miracle that was to occur that evening, that the sky over London should ever be so clear, but clear it was. The sky he saw was as clear as the sky over an Arizona desert or the mountains of Hawaii, and it was crowded with sharp twinkling stars. Some were brilliant, many were faint, but

all came thrusting within his field of view in untold numbers. Fainter yet he could see the dim, cloudy patches of distant galaxies (this really was a clear night); and as he watched, the bright trail of a shooting star flashed across his vision. For a time he was awed by this vision and briefly wondered what could be the meaning, what the nature, of all this marvelous celestial sight.

Soon enough, however, he came to his normal senses. He mentally shook himself and repeated inwardly that it was enough for a man to understand his own business, though he felt more than ever that he was in need of a vacation from his. "Humbug!" said Scrooge and let himself in through the front door.

Within, his apartment was warm and inviting and expensively decorated in soothing colors. Impelled by a residual nervousness he walked through its various rooms to see that all was right. Sitting-room, bedroom, kitchenette, all as they should be. Nobody was under the table, nobody under the sofa. He returned to the sitting-room, poured himself a large drink from his well-stocked cabinet and listened to some soothing music on his hi-fi system.

As he sat, lulled by the comfort of his deep armchair and beginning to relax, he toyed with his home computer. Abruptly it beeped and a message from his diary system appeared on the screen:

Scrooge stared at the screen in amazement; he had never left such a note! It was with a strange inexplicable dread that he heard at that moment his telephone bell beginning to ring and the answering machine switching on, relaying aloud the message it was receiving. He recognized, with no doubt whatever, the long-familiar voice of his erstwhile partner, Kevin Marley, delivering the unsettling message "Hello, Scrooge, be with you in a moment."

There was a clanking noise deep down below, as if some person were dragging a heavy chain over the central heating ducts in the basement. The heavy basement door flew open with a booming sound, and then he heard the noise much louder, on the floors below; then coming up the stairs; then coming straight toward his door.

"It's humbug still!" said Scrooge. "Such things have no place in the real world."

Without a pause it came on through the heavy door of his apartment and passed into the room before his eyes. Almost anticlimactic in its conventionality, the ghost of Kevin Marley appeared before him, dressed in his Italian shoes and heavy suede car coat but completely transparent for all that. Scrooge had always known his partner to be a particularly devious individual, but now the most naive client would be able to see right through him.

The chain was clasped about his middle. It was long and wound about him like a tail, and it was made (for Scrooge examined it closely) of mobile phones, filofaxes, and bankbooks. There was a liberal sprinkling of credit cards, all were gold cards of course, and on close inspection they appeared to be made of real gold and inordinately heavy.

Scrooge looked at the phantom through and through and saw it standing before him, but he did not believe it even now. Instead he saw before him the convincing evidence of overwork and hallucination. "How now," said Scrooge quite calmly. "What do you want with me? Whoever or whatever you may be."

"In life I was your partner Kevin Marley, but you don't believe in me. You do not believe in much, do you, Scrooge?"

"I do not," replied Scrooge. "I believe in my own concerns and little else. It is enough for a man to understand his own business and look to his own advantage, that is what I always say. My own business occupies me constantly, as I must live in the real world."

"In life I was as blind as you," replied the phantom. "I worked ceaselessly toward my own advantage as I perceived it and had no time for an understanding of the world, of creation, of life. Do not ask me what I am now, my dear Scrooge! There is no way in which you could understand, you who have no real understanding even of the world in which you live. Your world, the real world, the world inhabited by Mr. Scrooge and all humanity—what do you really know about it? How can you bear to live your life on the surface of this spinning globe which rushes through the immensity of space and not seek to understand what you may of the universe in which you live?"

The cold, hollow voice of the specter had become warmer and more emphatic, and it spoke with some emotion. Now it calmed it-

self and continued. "I have come to awake in you some understanding of the world; to show you something of the wonder and
the marvel of the real world, the world which you inhabit along with
every living thing. I have come to share with you a knowledge of
the physical world about you. You will hear much that will surprise
you, I think, and yet you will be told nothing that is not a part of
the knowledge you could have read about if you had wished. There
will be no special revelations. All that you will hear is already known
to your science, the science which you dismissed so glibly earlier to-
day.

"You will be visited," he continued, "three times tonight by Spirits who will reveal to you something of the world and its reality."

"I think I would rather not, if you don't mind," responded
Scrooge. "I am never at my best the next day if I have a disturbed
night. I tell you what, how would it be if I took out a subscription
to *Scientific American* instead? I do not really have the time or the
energy for this sort of thing."

"The first will come to you tomorrow when the bell tolls one,"
continued his visitor, paying no attention whatever to his remarks.
"The others will follow in due course thereafter. They will talk to
you of energy and of time and of reality in the physical world."

Scrooge gazed in dismay as the apparition began to walk backward from him, moving again to the closed door through which it
had entered. "Wait!" cried Scrooge. He was not overanxious to continue his meeting with this unsettling and uninvited visitor, but he
wished to know more of what was to come. "Wait! Can you not
say who or what is to visit me, as you foretell?"

The phantom paused just short of the door. "Your first visit will
be from the Mistress of the World and her Shadow. They will speak
to you of the science of times past, of ideas developed in previous
centuries. Listen well!"

Without further pause the ghost faded through the door by which
he had entered, and again the room was normal. Comfortable and
attractively furnished, it was just as it always was when he came
home to relax. "Humbug!" said Scrooge, but he said it quietly. He
felt to no small degree unsettled by his experience and was sure that
he would have little rest that night, even without supernatural visitors. He was consequently rather surprised to discover that he was
in fact feeling very sleepy as he undressed and got into his bed,
preparing to await whatever might come.

❧
The First Visitation

In which Scrooge is visited by Ghosts of Science Past, the Spirits of Energy and of Entropy. These shades would have felt at home in the time of Scrooge's famous ancestor, back in the reign of Queen Victoria, and they talk to him of science from that period.

He is shown how energy is the basic currency of the physical world. He is told that the total amount present in the universe does not change, so that energy is conserved, but that nonetheless there is great variation because energy may be transformed from one form to another. A falling object will show the conversion of energy from the potential of gravity to the kinetic energy of motion and finally to heat, but at all stages of the process the total amount of energy, of one form or another, is always the same.

He is shown also that, although the total amount of energy may be constant, the amount available for useful work is not. In any process the amount of available energy becomes less and entropy increases. More and more energy is tied up in heat, in the invisible random motion of atoms, until eventually all the energy in the universe will be randomized in this way to give the final Heat Death.

CHAPTER 1

The Mistress of the World

Although Scrooge would have sworn that there was no possibility of his sleeping that night, he nonetheless found himself starting suddenly awake from a fretful doze. What had awakened him? He looked around his bedroom to see what he might detect.

Now, being prepared for almost anything, he was not by any means prepared for nothing, and when no shade or shape became apparent he

was taken with a violent fit of trembling, but nothing came. He lay awake and five minutes, ten minutes, a quarter of an hour went by, yet nothing came. "Humbug!" he muttered to himself as he lay upon

11

his bed in the center of a clear core of light that streamed upon it and which, being only light, was more alarming than a dozen ghosts as he was powerless to make out what it meant. At last he began to think that the source of this ghostly light might be in the other room, from whence indeed it did seem to shine. With this idea now firmly in his mind he put on the silk dressing gown that lay by the bed and shuffled in his slippers to the door.

The moment his hand was upon the lock a strange voice called him by name and with great authority bade him enter. He obeyed.

In state upon an upright chair in the middle of his room sat a monumental figure, statuesque and glorious to see. It had the form of a mature woman of most imposing aspect and of more than human size. Her dress was imperial, and she projected an unquestionable authority of manner. Combined with her superhuman size, her imperial aspect suggested an idealized statue of Queen Victoria.

Scrooge was strongly reminded of some of the most difficult of his clients. On her head she wore a jeweled tiara which glowed and sparkled blindingly in the light of a bright clear jet of cold flame which sprang from the crown of her head and by which all else was visible. The light was twisted and distorted by the multifaceted gems in her tiara to give a moving, shifting array of strange shadows and brilliant spots of light upon the walls of Scrooge's living room.

"Come," she said inclining her head toward Scrooge. Without any conscious volition on his part, that one word compelled him fully into the room.

"Who and what are you," he demanded of her. His manner was perhaps a little less than civil, but he was still flustered by his sudden awakening and not a little put out by discovering this awesome invader of his property.

"I am the Mistress of the World," was her confident reply. "I am Energy, the very core and essence of everything that is and does and comes to pass upon this globe that you inhabit and indeed upon any other. Come, Man, and learn from me about energy, and you will learn much of the workings of this world."

"Indeed!" responded Scrooge, who had always found that it was essential to take a firm line with people of this sort. "I know that energy is important, and I am aware that the energy industries can be a very good investment, but why should it be quite as important as you say?"

"What is it," she countered, "that governs every activity in the world? What is it that decides what may and what may not be done, what is possible and what is completely beyond the possibility of achievement?"

"That is easy," replied Scrooge firmly. "It is money, of course. Money will determine what can and cannot be done. If you have money enough you can do anything; if you do not have money you are nothing. That is how things are in the real world, I am afraid."

"Not so!" cried the self-proclaimed Mistress emphatically. She rose from her chair to reveal that she overtopped Scrooge by a full head. Her blazing aura flamed up and dazzled Scrooge's eyes. "Not so. Does your money make the ocean's water rise into the sky as clouds and fall as rain upon the earth? Does it make an avalanche fall down the mountain's side or control the tides which surge upon the ocean's shores? Does it keep your world in its course about the Sun or carry the power to support life from that same Sun to all the creatures which live upon the earth? I think not! Your world has existed for a long time, little man, and many things have happened upon it long, long before your money existed, yes, and will continue to happen long after your money is no more. Money may be important to you humans, and indeed the love of it does sometimes cause you to bring destruction to the other creatures that share your world. It is of little moment to the world itself, to the real world on whose surface you exist.

"Come now," she continued more amiably. "What do you know of energy? Do you know its nature?"

"Certainly," replied Scrooge, who was determined not to be browbeaten by this domineering woman. "Energy is what we need to heat our homes and drive our vehicles. It is something that we get from out of the ground or perhaps directly from the Sun. As it is used up we must search for more sources to sustain our society."

"There is some truth in what you say. There is some truth, but there is untruth also. When I said that nothing happens upon the earth that does not involve and depend upon energy and the exchange of energy, I told you nothing but the strictest truth. Listen to me, Man, while I tell you about energy and the world. Energy is the essence of movement. It is the calm constant in the unending dance of the atoms. It is to be found in everything that moves or suffers change, but in this changing world it is itself always constant, however much it may change its outward form.

"Consider motion," she continued, half turning where she stood so that she faced toward a blank wall. "See what movement entails," she went on, leaning forward slightly in her chair and staring at the wall. Scrooge followed her gaze and observed the chaotic play of lights and shadows gradually coalesce and begin to form shapes that were, tantalizingly, almost identifiable. Abruptly the pattern sharpened into clarity, producing upon the wall a moving picture of brightly clad ice-skaters swooping to and fro across a floor of ice. It was not quite as sharp a picture as Scrooge could get on his widescreen high resolution television, but it was impressive nonetheless.

"There you have movement," stated the Mistress, rather unnecessarily in his opinion. "And what is the most obvious aspect of what you see? It is velocity, the speed with which the skaters travel over the ice," she continued, without giving him an opportunity to answer her question. "If speed is the most obvious property of a body in motion, is it also the most important, the most basic aspect of movement? It may seem obvious that it is, but how may we decide what is the true intensity of a body's motion?" Again she supplied her answer before Scrooge had time to make any reply. "We can observe what happens when motion is shared, when two moving objects collide with one another."

The picture upon the wall, which had been showing a great number of skaters as seen from a distance, zoomed in to focus upon one who was moving slowly, resting before her next burst of rapid skating. As they watched, another skater rushed carelessly across the scene and collided with the girl. She was propelled backward along the direction in which the newcomer had been moving, while he continued forward more slowly than before. It was as well that the images were silent, to judge from the expressions on their two faces. "There you see how the faster body has caused the slower to partake of his motion so that now she moves in the direction of his original velocity rather than her own. But is it only the speed which is relevant? Watch again!"

Once again, the phantom vision focused on a single figure that was moving more slowly than the surrounding skaters. This time it was a man of enormous girth, skating ponderously on blades which were dwarfed by his huge bulk. As before, a second figure raced into view. This was a slim youth who was looking over his shoulder at his admiring girlfriend as he skimmed forward with supple elegance, straight into the other skater. He rebounded at speed, windmilling his arms to retain his balance while the other continued on his way with little apparent change in his slow forward pace.

"There you saw a collision between two objects of very different mass, and the greater mass has dominated, even though he had the lower velocity. Speed is not necessarily more significant than the mass of the objects involved. In fact, it is the product of the two, the mass multiplied by the velocity, which is the critical quantity; a quantity that has the name *momentum*. The total amount of momentum does not change from one time to another. Colliding bodies may change the way in which the momentum is shared between them, as you saw when the moving skater gave some of his momentum to the girl who had been almost stationary, but the total amount of momentum present is always the same.

"Momentum is notable in that it is a conserved quantity. The sum total will never be more or less than it is now. Such constancy shows that momentum is in some way more significant than speed alone. It is in some way more bound up with the basic fabric of the universe than is mere velocity, which may increase or decrease quite arbitrarily.

"Although the *total* momentum possessed by any group of objects must stay the same, this amount may be shared in different ways among the members of the group as they interact and collide with one another. The total must remain constant, but at some time the individual objects may all have high momenta, while at another time they have very little. This variation is possible because momentum has a direction and as a consequence large momenta in opposite directions will cancel one another. Observe!"

Now the scene focused on a girl skating quickly toward the right, then the scene shifted and settled on another similar girl skating with equal determination toward the left. The viewpoint drew back to include them both, and Scrooge could see that they seemed to be on a collision course. They must have seen it too, but neither would give way to the other.

"There you see two bodies, both with quite a large momentum. Each has a momentum as great as the other's, but they are opposite in direction," said the Spirit. "The total momentum for the two together is the sum of a plus and a minus so the total is zero. Overall, taken as a pair, they have no net momentum at all."

The viewpoint zoomed in upon the two girls as they sped closer together, each holding her path until it was too late and they collided head on. They rebounded to stand face to face. Both were now stationary on the ice though admittedly their mouths were quite active. "There you see how both have come to rest, neither now has any momentum, but the total is still conserved because the total was already zero beforehand."

⟪ Momentum ⟫

Momentum is one of the *dynamical quantities* which has been found to be useful when describing the motion of objects. At low (nonrelativistic) velocities the momentum p of an object is given by the product of the object's mass *m* and its velocity *v*.

$$p = mv$$

This equation shows that momentum depends equally on the velocity *and* the mass of the object.

Velocity has a direction as well as size. It is what is called a vector quantity. A particle moving from left to right has minus the momentum of a similar particle moving from right to left with the same speed.

Momentum is important because the **total amount is constant:** It is *conserved.*

However, if particles are moving in opposite directions they may each have a large amount of momentum, but the total may be small or even zero.

$$mv + (-mv) = mv - mv = 0$$

"But surely the momentum changes when one of them begins to build up speed again," protested Scrooge. Despite his initial displeasure at the uninvited invasion of his home, he had now become quite caught up in the problem. "As one skater skates away, her momentum must be increasing because her mass will not alter and no one else is involved to balance this momentum. The total must change."

"But that is not so!" replied the Mistress. "There may not be another person involved, but a skater cannot increase speed without pushing against the ice. As she pushes herself forward so she pushes the ice backward and her forward momentum is balanced by a backward momentum given to the ice."

"But the ice will be attached to the ground!" protested Scrooge.

"Certainly it is. In consequence, the momentum of the ground must change also! If you begin to run toward the east the speed of rotation of the earth will slow down, and if you run toward the west it speeds up. Of course the mass of the earth is much greater than your mass so the change in the earth's speed is not perceptible, but it *will* change. You can see the effect more clearly when the two masses are not quite so different."

She gestured toward the wall with a substantial arm, and as Scrooge looked the picture changed once more. This time it changed not only in content but also in style. The scene he saw was shown in flat areas of color, each separated by a heavy black line. He saw a skateboard sitting on a highly polished floor with its front wheels close to the edge of a long steep staircase which was shown in dizzying perspective as it dropped away into the distance. On the skateboard was a small gray mouse with round ears and large appealing eyes. Scrooge imagined he had recently seen this same scene, or one much like it, on a childrens' television program.

The mouse began to scamper away from the terrifying drop and, as he did, the board beneath his feet rolled in the other direction so that its front wheels teetered over the edge. Abruptly the mouse stopped, his eyebrows rising to hover some distance above his head and a large question mark popping into existence over him. He scurried back the way he had come, and the board withdrew slightly from the brink. Cautiously he crept on tiptoe toward safety, but as he did so the board edged closer to the drop. Noticing this he stopped at the same position he had reached before, looked out toward Scrooge and shrugged helplessly.

"You see how the lighter body moves farther and faster than the heavier one, but both must move if they are free to. If one of them is fixed to something else, even if it be to the earth itself, then that must move to conserve momentum. In all cases momentum will be conserved as I have told you. So you see, though momentum is a less evident property than velocity, it is truly more significant in that it is a *conserved quantity*, and there are simple rules about the amount of it which may be present. If momentum is fundamentally important, even though its importance may not seem immediately self-evident, how much more so is energy? Energy is a conserved quantity and an even more important one."

Scrooge had temporarily forgotten that he was to be instructed about energy. Now apparently the preamble was over.

"Everything which moves has energy," continued the Mistress of the World, settling herself comfortably upon her chair. It was obvious to Scrooge that this explanation was going to take some time.

"Any moving body has a certain amount of energy, which a stationary one does not," she continued. "This energy is solely due to the motion and is called *kinetic energy*. It is rather different from velocity or even from momentum. Like momentum it depends on the *mass* of the object in motion, but unlike momentum it does not depend upon the *direction* in which it is moving. You saw that the momentum of a body moving from left to right is opposite to the momentum of a similar object moving from right to left, and that if two such movements were combined the total momentum would be zero.

"Energy is different. Energy does not depend on the direction of motion. The energies of two moving objects will never cancel, they can only add. Energy depends on the square of the velocity, you see." She paused to explain: "That is to say, the velocity is multiplied by itself. Any negative sign is thus canceled by another, and the result is positive. As a consequence, adding the kinetic energies of any two particles cannot reduce the total."

"But does that not mean that the moving skaters must have had kinetic energy?" queried Scrooge, who felt that he had spotted an obvious flaw in what he was hearing.

"Certainly it does."

"And should not the total amount of energy they possess be constant, from what you say?" pursued Scrooge, who was determined that there should be no equivocation.

"It should. It is."

"How then did it happen that before they collided the two girls were moving quickly, and so both must have had this kinetic energy you describe, but afterward they had both stopped and so had none. Does this not mean that this energy is not conserved but is in fact transient and easily lost, being present at one time and the next moment gone?"

"No, it does not. Admittedly the amount of *kinetic* energy had changed in that instance, but kinetic energy is not the only form that energy can take. The universal importance of energy arises because it has many aspects. Watch closely."

At a gesture from the Mistress, two tennis balls that had been sitting on a table rose into the air and began careering around the room, bouncing from walls and furniture. They headed directly toward one another and collided before Scrooge's eyes. As they collided he saw that they lost their spherical appearance, and as they came momentarily to rest they were both squashed into flattened pancake shapes. In a moment they had rebounded to their original form and shot off, apparently as quickly as before.

"There you saw the kinetic energy of those tennis balls converted briefly into another form of energy, which stored the work that they had to do to and so distort their shape. This energy was stored within the balls and released when they returned to their previous shape, appearing once again in the form of kinetic energy as they sped apart.

"Energy is very closely related to work; indeed it is formally defined as the capacity for doing work." Scrooge perked up when he heard this, for he was a great believer in work, hard persevering work, and the thought that work was of fundamental importance to the nature of the universe appealed to him mightily.

"This is of course work in the physical sense," explained the Spirit. "It is defined to be movement against the resistance of an opposing force. In these terms a man who shovels dirt out of a hole in the road against the downward drag of gravity is doing work, but I am afraid that you, sitting all day at your desk, are not!"

Scrooge felt mortally offended by this comment, but before he could formulate a sufficiently cutting reply she had swept on regardless. "The concepts of work and of force have been very important in people's attempts to understand physical dynamics: the way in which things move around. If you can offer no resistance, a force will push you where it will, or it may oppose and prevent you from reaching the place to which you were going. Opposing the effect of such forces requires that you do work in the physical sense, and the possession of energy allows you to do this. Energy is in effect the currency of the physical world. If you have energy it allows you to do many things, to go where you will, despite opposing forces; if you have not, then you are quite at the mercy of external influences.

⟪ KINETIC ENERGY ⟫

Kinetic energy is another dynamical quantity which is useful when trying to understand motion. The kinetic energy E of an object with mass *m* and speed *v* is given by

$$E = \frac{1}{2}mv^2$$

As with momentum, this depends on both mass and velocity, but now the effect of the velocity is greater as it is *squared*. This means also that the kinetic energy is positive whatever the direction of motion. Plus or minus both give plus when squared. Adding another particle to a set of moving particles will always *increase* the total kinetic energy.

Energy is a conserved quantity. The total energy of any system is constant: a very important result which is also known as the *first law of thermodynamics*.

The *kinetic* energy need not be constant though. There are other forms of energy, and one variety may convert to another. It is the total amount of all the types which is constant.

"Not all energy is manifest as movement," the Spirit continued. "Energy may be stored in many forms, as potential energy as well as kinetic. Kinetic energy is the form that is easy to see because its possessor rushes frantically around. Potential energy is quiet and hidden, but where there is energy it may be converted back to movement. You can have potential energy in many different forms: as chemical energy in the food you eat and the petrol that you burn in your car, or as stress energy, which is stored in a distorted solid, like the tennis balls which you saw colliding. The most obvious form is gravitational energy. The potential energy of any massive body increases as it rises from the surface of the earth. You have to do work when you lift a heavy weight, and the energy provided by this work is fed into the increasing potential energy of the weight. Release the weight and it will fall, converting this potential energy to kinetic energy as it plunges downward with increasing speed.

"You may see the cycle of conversion repeating over there," she continued, rising from her chair and moving toward him. As she approached him, Scrooge was overwhelmed by her monumental presence. She overtopped him by a full head and made him feel childlike in comparison, a feeling he did not appreciate. She seized his arm and turned him toward an antique grandfather clock which his interior decorator had deemed appropriate for the room. Behind the glass door a massive pendulum could be seen, swinging slowly to and fro.

"There you see kinetic energy turned to potential and back again to kinetic, the cycle repeating again and again as your clock marks off the seconds of the day. At the bottom of its swing the pendulum bob moves quickly and has kinetic energy. As it rises it slows until its kinetic energy is quite gone, but now it is higher and its energy is stored as the potential energy of gravity. Down it falls from this temporary height and the potential energy converts back to kinetic. Again the bob is moving quickly. It is moving in the opposite direction to what it had before, but this is no matter as kinetic energy does not depend at all upon the direction of its motion. Next it will rise and come to rest at the other extremity of its swing. Then again it will fall; rise and fall: rise and fall, with its energy forever cycling between the kinetic and potential forms. In this clock of yours the cycle will not actually last forever, or indeed for very long if you do not wind it, because on each swing the pendulum loses a little of its energy to the surroundings from the effects of friction and the resistance of the air. Despite this, a large pendulum that carries a lot of energy can go on swinging for a long time."

Scrooge was not sure that he was convinced. It seemed to him that new rules were being invented just to explain away every discrepancy that he suggested. "How can you say so confidently that energy is conserved when it changes so often from one thing to another? You may say that kinetic energy vanishes and something else appears, but have you reason to say that they are in any way the same thing?"

ᗍᗌ CONSERVATION OF ENERGY ᗍᗌ

This is one of the great general statements of physics. The total amount of energy in any isolated system does not change: It is *conserved.*

This is not to say that kinetic energy must remain constant, as one form of energy may *convert* to another. The kinetic energy of a rising ball may *convert* to gravitational potential energy as the ball rises and slows. Some of the kinetic energy in a spinning dynamo may *convert* to electrical energy, while some may be *converted* by friction at the bearings into thermal energy: The material gets hotter.

In this process energy *is* conserved. The rate of exchange for the conversions is fixed, and the amount of energy distributed among all the forms is constant. At the end of Chapter 10 we shall see that every form of energy is in some way the energy of particles, so the different forms are in fact the same thing.

"You can see that they are equivalent, since the various forms of energy convert to one another and always produce the same amount of each form. There is no arbitrariness. Your pendulum has the same speed at its lowest point on each swing, it rises to the same height at each extreme. The different forms convert to one another with strict equivalence, like some conversion between currencies that takes place with no commission for the transaction so that you can get back all that you had originally. However many conversions take place, no energy is ever lost. Some may be lost to you and your purposes, as when the pendulum of your clock loses energy through friction, and if you like you may think of this as a commission on the transaction; but it always goes somewhere else. It does not simply vanish. The bookkeeping is exact, and Nature's currency of energy does not suffer from inflation.

"Come!" she said, the word as commanding as when she had first spoken it to Scrooge. "Come with me, Man, and see how the play of energy features in this your city."

Preparatory to venturing outdoors she picked up, from where it had lain beside her chair, a hat shaped much like a great candle extinguisher or dunce's cap and, clutching this under one arm, she drew Scrooge forward with the other. The pressure was gentle, but he was no more able to resist than he could resist the turning of the earth itself. Around them the room vanished, and they stood in the open before a vast hall with blank walls soaring to the sky. To one side stood strange gray towers from which poured dense clouds of steam, rising and mingling with the gray sky overhead. To the other side he could see a great spreading hill, coal black in color and fed from rows of railway wagons marshaled to one side.

"This is an electricity generating station. Much of the energy that pours through the life of the city is carried by the medium of electricity. Electricity is easily carried from house to house by wires beneath the streets; it is convenient and versatile in its application, but it must first be created by converting other forms of energy." Scrooge followed the Spirit as she led him through the huge building. He saw where coal was fed into large furnaces, burning to convert the chemical energy stored within its substance into heat. He saw how the heat generated steam in large boilers and how the rushing steam turned great turbines, so the heat was converted to the kinetic energy of the spinning shafts. The shafts led in turn to other great machines, generators in which the motion of their cores produced electrical energy, which was sent forth to the city.

As they left the building, Scrooge commented on the tall gray towers with their plumes of steam. He was told how, in each step of the process by which the energy locked within the coal was coaxed toward the form of electricity, some energy slipped away as heat. Some stages released more heat than others, but in every step heat was produced, do what one might to prevent it. This heat was removed by boiling water in the heat-exchanging towers and releasing the steam into the atmosphere.

From the power station they followed high wires strung on pylons that carried the electricity into the city, and there they traced the wires that traveled beneath the pavement, branching off to enter every office, shop, or dwelling. They saw how electricity could become the energy in light, from the multitude of tall chandeliers in a concert hall to the tiny glow of a night light by a child's bed. They saw energy convert to heat, this time by intent, in bed-sitting rooms where students worked beside the glowing filament of an electric fire.

Within people's homes they saw electricity used for many purposes. For the handyman, electricity would provide the kinetic energy of his spinning drill, for the cook it gave the heat in an oven and the more subtle current in the circuit that switched that oven on and off at the times required. To while away leisure, electricity gave light from the television in the corner of the room. It gave sound from the speakers of a stereo system. It gave, well, I can scarcely say what it gave from the children's video games. In all of this, the energy delivered by the electricity was converted to many forms—heat, light, sound, or motion—but in every case there was some heat.

Not all of the energy used came in the form of electricity. In many homes the heating was through the burning of oil, though electricity controlled this. Out in the streets people moved around in cars and trucks which found the kinetic energy of their motion from the burning of petrol. In all of the activity about him, even in the motion of the people's own bodies, which they fueled with the energy from food, Scrooge saw a complex dance of energy, converting from form to form.

And now, without a word of warning from the Spirit, they rose and sped over the city, high above all the activity of street and house. Together they sped on, on—until, being far away from the teeming city, they stood upon an open countryside that spread before them under a lowering sky, dull, flat, and devoid of any significant feature. Stunted trees and scrub filled a shallow valley which was ringed by low hills, on one of which they stood.

"You have seen how energy moves through the various activities of your city," said the Mistress. "Now observe the devastating energy which Nature can release." So saying, she again seized Scrooge, not this time by the arm but in some way she enfolded his awareness and, although his body remained standing at her side on the hill top, he felt himself flung up into the sky. Through the gray clouds he rose until he burst out above them and saw the sunlight upon the fluffy carpet of their upper surface. Up, up he went, rising so far above the clouds that he began to see the curvature of the earth's surface. Still he soared, higher and higher, while the earth became a ball beneath him, the familiar continents upon its surface vaguely glimpsed through a brilliant cover of swirling cloud.

Farther and farther he traveled: The earth shrank to a tiny size, and he receded into the cold darkness of space. For a brief moment he passed close to the half-lit sphere of the Moon, swinging in its orbit around the earth, but soon that too was left behind and he was alone in the dark emptiness of space. He felt totally isolated and lost. Was he still moving away from the earth? There was no way to tell, no point of reference, no sensation even of the passage of time.

Eventually he became aware that he was not completely alone in the void. To one side he saw a boulder. Nothing more than that. A cold silent lump of rock, illuminated by the distant Sun and drifting gently through the wastes of space. Whether it was small and close at hand or very large and very far away, there was really no way that he could say, but it was hard to imagine anything that showed less sign of energy or activity of any sort.

For an unknown time he seemed to drift alongside this dull, lifeless stone with no feeling of any movement, with scarcely any feeling of his own existence. Eventually, after a time, which might have been hours or centuries for all he knew, he saw something appear in the darkness ahead. It expanded until he could recognize it as the shining ball of the earth that he had left. Steadily it grew and grew in size until its surface was all that he could see before him, the pattern of land and sea just visible beneath the thick white swirls of cloud.

As he tumbled downward he looked across at his faithful dull companion and imagined that he saw a faint flickering tail begin to form behind it. More time passed and this tail became clearly evident, with streams of gas and vapor trailing behind as the rock en-

countered the first thin verges of the earth's atmosphere. This turbulent tail increased by the minute as the falling boulder was drawn ever faster toward the planet below. The foremost surface of the stone now began sullenly to glow, like one face of a dying coal.

Brighter and brighter waxed the glow, more and more intense was the tail of flaming gas which streamed behind as it fell ever faster in the grip of the earth's gravity. Soon it was a blinding ball of gas, its center lost within the flames and even the solid rock converting to gas in the intensity of the heat produced. If Scrooge had been present in the flesh he would have been instantly blinded, though this would not have inconvenienced him so much as you

might imagine, since he would thereafter have been burnt to a wisp of vapor upon the instant.

Incorporeal and indestructible, he hurtled toward the surface of the land in strange companionship with this fiery meteor, for so he realized it to be. The gravitational energy released when the drifting rock had chanced to fall upon the earth was being converted into the kinetic energy of its present headlong flight, while the atmosphere's attempt to resist this invasion of its realm and slow the too-rapid fall was bleeding energy from its motion and converting it to this furnace-heat.

A much smaller stone could not have survived this journey and would soon have shed all its mass in its fiery passage through the air, burning itself up in the higher reaches of the atmosphere. This meteor, however, was no tiny pebble, this was a considerable rock with mass enough to survive the passage to the earth's surface and on this surface in due course, hidden within its sheath of blazing plasma, the meteor finally struck.

Even as it approached the ground the incredibly hot gases in its sheath heated with an explosive suddenness the surrounding air. The air expanded manyfold and burst upon the countryside around a flaming wind which tore across the stunted trees, flattening each one to the ground and as it lay converting it instantly to a flaming torch.

The meteor's fall had been resisted stubbornly all of the way by the thickening mantle of the earth's atmosphere, and this resistance had converted much of its kinetic energy to heat, boiling away the very rock so that it steadily shrank in size. But, when all is said, the air was only air. When the meteor finally struck the ground it discovered real resistance to its passage. In a short distance that monstrous kinetic energy had been converted to heat and all of the mass of the falling rock to blazing vapor, together with a sizable chunk of the earth's surface.

Not all of the energy was converted to heat. Some went into compressing the substance of the surrounding rock, which then rebounded violently, sending out waves, like a stone dropped into a pond. These waves raced through the solid ground, and as they passed they tossed aloft anything which was on the surface. From the region where the meteor had struck, huge quantities of dirt and dust were flung into the air, blotting out the light of the Sun.

Scrooge found himself again in his own body and might have expected to be summarily destroyed by the violence that surrounded them, save that in some way the Spirit cast a shield about him which spared him bodily harm. At the same time, since the land was darkened by the dense pall of dust, the jet of light upon her head surged up to illuminate his surroundings. She turned to address Scrooge once more.

"There you have seen how the gravitational potential energy which a mere stone releases when it falls upon the earth may convert to other forms. Initially it becomes kinetic energy as the stone falls ever faster but, since it acts against the resistance of the air, much of this energy is converted to other forms. Much of it goes to heat, some goes to chemical energy as it breaks the bonds that hold together the atoms in the stone, some goes to compressing the solid ground, and some to kinetic energy again as it moves great volumes of air in the howling winds or of earth in the heaving land."

They stood for some time observing the burning winds, themselves the only objects that remained unscathed in the midst of this cataclysm of nature. Gradually the moaning winds died down, the flaming trees burnt themselves out, and the shuddering earth grew still. As more time passed, the ashes cooled and the glowing pit of the meteor's crater faded to dull, cold rock. As it had been at the beginning, the scene before them was once more dull and undramatic; no sign of the recent tumultuous events remained save for the cold central pit and the twisted bones of motionless trees under the dark, dust-laden sky.

"Why, all is now as calm as it was before. Where has all that energy gone to?" demanded Scrooge. "Where are the winds, the furnace heat, the shaking earth? It has all gone! Every single form of energy that you have named has faded away. Despite everything you have said about its conservation, the energy has simply vanished! I will be deceived no further," he cried out in sudden fury and, seizing from the Mistress her cap, which was so conveniently shaped like an extinguisher, raised it at arm's length above her head and pressed it down upon the jet of light.

The Spirit dropped beneath it and seemed to shrink so that the extinguisher covered her whole form; and as Scrooge pressed it down, the light streamed from under it in an unbroken flood upon the ground. He put his full weight upon it and forced it down with all his might, finally cutting off the ring of brilliance between the cap's rim and the earth. Now the Visitation's brilliance was finally dimmed, but still some light escaped because the cap did not block it completely, allowing a diffuse glow to filter through its sides. This light was insufficient to illuminate the scene in any useful way, and all around him Scrooge could see nothing but a diffuse grayness. As he watched, a patch of this grayness took a form but marginally darker than the rest and strode toward him.

$\mathcal{E}$

CHAPTER 2

The Shadow of Entropy

Scrooge strained his eyes to peer through the deep surrounding gloom. He tried to be certain whether there was in truth a denser patch of shadow which was moving toward him; and if it was not a trick of his imagination, whether he could make out some detail within it. The haze of dust and cloud which hung about the scene began to clear, and he saw the oncoming shadow resolve itself into the outline of a human figure. Within this outline, however, all detail still remained unclear. The shadow advanced steadily toward Scrooge and at last stood directly before him. By that

time much of the dust had settled, and he could once again see the scene about him. He saw the blasted trees and the distant undergrowth, but this figure, although so much closer to him, remained nothing more than a silhouette filled with an impenetrable blur. The figure's outline was far from clear but suggested the outline of a tall and rather substantial woman, dressed in a flowing gown and wearing some sort of headdress. It looked, in short, like a shadow of the Mistress herself.

"You asked where the energy has gone." It took Scrooge some moments to realize that the figure had in fact addressed him, so dull was its speech. Its voice was a distillation of all the tedious and long-forgotten lessons through which Scrooge had been forced to endure as a child. Its speech had the clarity of a distant announcement heard on an echoing station platform and the impact of a series of political broadcasts for a rival party. It was, in short, a voice which would have been all too easy to ignore, save that in this scene of such recent devastation Scrooge was particularly receptive of any message.

"Who and what are you," he demanded of the dim shape. "Are you yet another of the Spirits who have been sent to inform me?"

"I am," came the blurred reply, "I am the Shadow of Entropy, a quantity that governs how energy is transferred. To many people the whole concept of entropy seems shadowy, although the effects of entropy are great and far-reaching. You have already seen something tonight of how the energy possessed by a system controls what *may* happen. Entropy further controls what is *likely* to happen, what the future *will* bring to pass."

"Very well," responded Scrooge. He could not say that he felt quite comfortable to be so accosted by passing Spirits, but he was becoming more accustomed to the idea. "If entropy is as important as you say, then say on. Proceed if you will and tell me then; what is entropy?"

"That is not a question which may be quickly answered," replied the Shadow rather evasively. "It is sometimes said that entropy is a measure of disorder, but I fear that such a description tells you little. In order to understand something of entropy we must first speak about *probability*.

"You asked where the energy has gone," she repeated abruptly. "It has not gone anywhere; it is still all around you. Energy is indeed conserved, just as you were told. There is exactly as much of it as there ever was, but you are no longer aware of its presence because it has been distributed evenly as heat. Every part of the surroundings is a little warmer than it was before, but only a very little because the available energy has been spread so thinly over such a wide area. Everything is now but a little warmer than it was before. The ground, the air, the trees, even you yourself, are all a little warmer. As they are *all* warmer each is still at much the same temperature as the others and so you are not aware that there has been any change at all.

"You do not notice that any change has taken place in your surroundings or see what effect the absorbed energy has produced because you look only at the outward seeming of things. You do not see into the heart of the material world because you cannot see the *atoms* of which it is composed. Look closely at the ground now and I shall help you to observe the nature of matter."

Scrooge looked down at the ground as he was bid, focusing on a bare patch of rock near his feet. This was a totally unremarkable rock, no different from hundreds of other rocks which he had seen and not worthy of any particular notice. As he looked intently at this patch of rock his view of it appeared somehow to become clearer than he would normally expect. The details became sharper, and the surface seemed to rush up toward him, the view expanding and sharpening as more and more surface features became visible. He saw large patches of colored lichen spotting its surface, and these grew and spread until they looked like some ground-hugging undergrowth. He saw an ant crawling over the rock and this also expanded vastly in his vision as he watched. Its armored head swung toward him, and its large compound eyes seemed to fix upon him the gaze of an inquiring hound before the insect slipped sideways out of his field of view.

Still the rocky surface expanded before him. A small crack in the stone grew and grew in his perception until it seemed as if he was hovering over a great rocky canyon. At its bottom he could see a lake, which was in reality a film of dew upon the rock. Down, down into the depths of this chasm he descended while cliff walls rushed past upon either side. Soon he approached the bottom and was among a throng of cliffs and pinnacles. These had a strangely regular aspect, as if they had been built by welding together a set of building blocks to form mounds with square edges and sharp cubical outcroppings.

"You are now seeing the first signs of geometrical regularity produced by the underlying crystal structure of the stone," remarked a patch of shadow behind one of the outcroppings. Scrooge realized that his companion had followed him into his expanded perception. "This regularity of the material is a reflection of the fact that the

atoms in the crystal fit together in a regular array. Soon your perception will have reached a level where you are aware of the atoms
themselves."

Sure enough, the rectangular forms enlarged and at the same
time grew less solid looking, resolving into apparently endless arrays of hazy spheres like a huge number of out-of-focus beach balls
stacked in erratic rows in some vast warehouse. Scrooge assumed
correctly that these spheres were atoms. He could make out no detail, but it was obvious that they were in motion, each one vibrat-
ing about its proper position in the array. Some vibrated violently,
others more gently, but all were in motion.

"There you see the presence of atoms within the material. Indeed the atoms *are* the material. Everything which you see in the
world around you is made from atoms. Huge numbers of atoms of
various types in various combinations make up all the different substances which you may encounter, and all of these atoms are in motion. In solids the atoms are anchored to fixed positions, as you see
here. They are unable to move far afield, and their movement is limited to a vibration to and fro.

"In liquids the case is different," she went on. Scrooge found his
view sliding sideways until he no longer saw the regular arrays of
atoms within the rock but saw instead those within the pool of water. Here also were the hazy spheres, and they were packed almost

as closely as they had been within the solid. They were not held to any particular position now but moved freely past one another. The individual atoms were moving in every possible direction, each with a different speed, weaving past one another and brushing one another aside like a dense crowd in a shopping mall.

"As you can see, the atoms in liquids are almost as closely packed as in a solid, but they can move around quite freely, slipping past one another as they move. In gases the atoms are much farther apart, and each can move for quite a long distance before it collides with another. When they do collide they bounce off, and as they do they transfer energy between them. This also you may see, as above the rock and the water there is the air, which is a gas. All you have to do is draw back a little to see the atoms within the air."

With no apparent volition on his part, Scrooge found that the region he could observe seemed to be withdrawing up out of the liquid, and in due course he was indeed looking at the air. Here were the same atoms, or at least similar ones. They were no longer in close contact but spread out far apart with a great deal of space between. Now that they were so well separated he could see that the atoms were not completely solitary but, rather, were moving around in pairs. "Molecules," the familiar voice murmured in his ear. "The atoms are grouped into molecules. There are not so very many different types of atom, but they tend to stick together in collections of atoms called molecules. The different combinations which they form give all the vast array of materials that you know. The molecules of air are rather simple, being nothing more than two oxygen atoms or two nitrogen atoms stuck together."

Scrooge regarded these groups of atoms and saw that they were moving in every direction imaginable and with a great variety of speeds, some dashing along rapidly while others ambled much more slowly on their way. Because they were so widely separated they could travel without hindrance in a straight line. At least they could do this for most of the time. Every now and then one would collide with another, and both would fly off in different directions. When this happened the slower molecule usually moved more quickly after the collision while the faster one lost some of its speed.

On occasion a rushing molecule would run into the ranked array of atoms at the boundary of the solid. When this happened, the molecule usually rebounded back into the gas, reversing its direction of flight and its momentum. This required that it transfer a balancing momentum to the solid, and through this continuing rain of molecules the gas exerted a pressure upon the solid.

◖ Kinetic Theory ◗

The kinetic theory explains many of the properties of substances, particularly gases, on the assumption that they are made of large numbers of small particles, atoms, or molecules, which rush round and bump into things.

The heat in gas is given by the sum of the kinetic energies of all the atoms. The mean kinetic energy for the individual atoms is a measure of the gas's temperature. When the atoms bounce off the walls of whatever is containing the gas, they push them outward and this gives the pressure of the gas. The kinetic theory explains the simple properties of gases very satisfactorily.

Scrooge noted that in all the different conditions in which he had been shown atoms and molecules they had in every case been fuzzy, with no detail visible. Though he could see that the atoms were there, he could see nothing at all of their nature. "Can you not further expand my view," he asked the Spirit. " It would take but a little more magnification and then I could see the nature of these atoms. At the moment I can in no wise make them out."

"No, that I cannot do," was his reply. "However much further your view were to be magnified, the atoms would appear no sharper. This fuzziness of which you complain is an essential part of them. One of my younger brothers will talk to you later of the quantum uncertainties on this scale of being.

"Even though you may not have seen them clearly, you cannot have failed to notice that in every case the atoms were in motion. There is kinetic energy involved in the motion of every atom within the material. This energy, the kinetic energy of all the countless moving atoms, is what you know as heat. The faster the atoms move, the more heat is stored and the higher the temperature of the substance.

"You will have noted that in general the atoms move differently from one another. Some move quickly, some relatively slowly. Some move in one direction, some in another. As they move they frequently collide, and in so doing they redistribute the energy among themselves. This movement is where the energy possessed by the meteor has gone. It has increased slightly the *average* energy of motion of all these myriad atoms which teem within the earth at your feet."

When the Shadow finished speaking Scrooge saw that the view presented to him had shifted back to the atoms within the solid, each one oscillating to a greater or lesser degree about its proper position within the material. He could see interminable arrays of the atoms extending in every direction, and every one of them was in motion. Clearly there was much energy involved in the motions of so many atoms.

Abruptly he experienced a wrenching change of viewpoint and found that he was again standing upon the devastated plane, looking with a new appreciation at the undistinguished patch of rock by his feet. To his normal sight the rock remained no more remarkable than any other rock he had seen before, and so he switched his gaze to the Shadow, who had come up close beside him. There was nothing remarkable to be seen in the Shadow either; in fact, Scrooge thought that she did look remarkably *like* nothing. He abandoned this thought as he realized that the Shadow was once again addressing him in her usual tedious tones.

"You will have observed that each one of the atoms that you saw in the solid was doing nothing very significant, merely swinging to and fro about a fixed position like the pendulum of a clock." As she said this, a large clock which looked exactly like the one from Scrooge's own apartment mysteriously appeared nearby, standing by the trunk of a burnt-out tree.

"You see here a pendulum. It is swinging back and forth as a pendulum does. Its motion is repeated again and again without any apparent difference. I ask you, why does it cycle like this? If you move such a pendulum from its lowest position, the position in which it will hang at rest, why should it start to swing in this way?"

"I had always understood that the force of gravity pulled it back down toward its lowest position," answered Scrooge promptly. He was determined to show that he was as well informed on most subjects as the next man.

"Ah yes, the force of gravity," murmured his companion. "Your answer is of course correct, as far as it goes. Yes, the force makes the pendulum move, indeed it does. But tell me then, what is a force?"

Scrooge was on the point of answering "Something that makes something move," but he changed his mind at the last minute. He felt somehow that this answer might not add very much to the discussion. Instead he switched to the offensive and challenged his companion "Very well, you obviously have some point to make, so you tell me. What is a force?"

"That is a good question, although perhaps a little similar to mine. It will help in the long run to think in terms of energy rather than forces. When you pull a pendulum to one side you raise the heavy bob at the end slightly above the level which it has when at rest. This increases the potential energy it gains from gravity. The farther anything is from the center of the earth, the greater is the potential energy it gains from the earth's gravity.

"The pendulum may not move very far from the center of the earth, but it does move a little. As the pendulum falls it loses this potential energy, which is converted to the kinetic energy of the moving pendulum. Any physicist can tell you that the force of gravity is simply equal to the rate at which potential energy changes. They are precisely the same thing. When we see something being moved by a force, we are equally seeing that Nature, for some reason, likes to decrease potential energy and to change it into kinetic energy."

"But that is not all that happens," protested Scrooge, who had been carefully following this rather strange-sounding argument. "The pendulum may, as you say, be losing potential energy and increasing its kinetic energy *until* it reaches the bottom of its swing, but after that it will rise again and the kinetic energy will change back to potential. That is quite the opposite of what you are saying."

"True, so it would seem. So it does, in fact. I should point out that you have exactly the same difficulty when you speak in terms of forces. Once the pendulum has reached its lowest point and begins to rise, it is moving *against* the force of gravity. The pendulum is forced to do this because it is *constrained*, its options are severely limited. It is fixed to a pivot and can only move along one path. When it is at the lowest point on this path it has its greatest kinetic energy, but it cannot just stop there. It is moving, and so it has momentum as well as energy. Because momentum is conserved it must keep on moving, and because it is connected to a pivot this means that it swings upward again. As it swings up, the pendulum gradually loses both kinetic energy and momentum. The kinetic energy is now converted back to potential energy because the pendulum is rising; there is no other option available to it. The momentum is in fact transferred to the pivot, and whatever is holding the pivot, because the total momentum remains conserved. This will sway to and fro in opposition to the pendulum, but the distance it moves will be tiny because it is so much the heavier.

"Whenever something speeds up, or slows down, or is deflected to one side, then indeed it may seem reasonable and useful to talk of a force that is acting on it. There are other circumstances where it becomes difficult to visualize that a force is at work. A flame that gives off light provides one example. Can you imagine that some sort of force is pushing the light out of an atom? That it is pushing out some sort of light which is already in the atom and need only be forced out of its hiding place. I think not. That is not a case where the notion of a force is at all helpful, and in such cases it makes little sense to talk about forces. I can confidently say that 'the Force *is not* always with you.' "

The Shadow began to stride up and down as she lectured Scrooge. It was a strange sight. The diffuse outline of his companion billowed and spread, as might the outline of a full skirt which swept around with each turn across the ground. Scrooge could not help but ask himself why all the personal features which his companion presumably possessed should be so hidden from him.

"Earlier, when you looked closely at the ground, you saw that it was made of atoms and that the behavior of each atom was quite simple. In many cases the atoms oscillate continuously around a central position, much like this pendulum. They may transfer their energy from one to another to give different overall states of motion for the entire collection. Where the energy may be so freely shared, then you can as readily produce *any* of the states that share out, in whatever way, the same total amount of energy. It is reasonable to assume that you are equally likely to find any one of these possible states as there is nothing to choose between them.

"Energy may be transferred from one atom to another. It might

just as readily be transferred back to the first. There is nothing in the idea of energy conservation to say *in which direction* energy may be converted. The interactions that happen between atoms are symmetrical and perfectly reversible. In the case of a pendulum swinging regularly to and fro you would be hard put to distinguish any distinctive difference between its past and future motions.

"All of the interactions and energy transfers that are going on between the individual atoms are reversible in this fashion. Energy is just as likely to be transferred one way as the other. There is absolutely nothing to choose. Let me show you a collision of two atoms."

Within the fuzzy mist of the Shadow's body, if body indeed she had, Scrooge saw a bright window appear. It looked rather like a small television screen and within it he saw the image of two hurtling atoms which crashed together and rebounded. On the rebound one was traveling visibly more quickly, the other more slowly, than before. Scrooge could not avoid the thought that the collision seemed strangely disappointing compared with the dramatic collisions between vehicles he saw so frequently on his television set.

"Now see another picture of the same collision," ordered his companion. Scrooge obediently watched. The speeds and directions of flight before and after the collision did appear to be a little different from the previous time, but not in any significant fashion. He felt there was nothing to choose between the two sequences. "Those sequences both showed the same collision, but one of them I played in reverse. Can you tell me which of the two views was the correct one?"

"No, I cannot," answered Scrooge. "There seems nothing to choose between them. In each view one atom collides with another, and that is all I see."

"You see well enough. There is no way that you could have told which sequence was in the correct order. The reversed sequence is just as likely, just as reasonable, as the correct one. When you look at one or two atoms there is nothing in their behavior to make a distinction between past and future. Assuredly, they change and behave differently at different times, but there is no progression, no development, no history. In the collision of two atoms you see no *arrow of time*, nothing which points the way uniquely toward the future."

The Shadow strode over and stood beside the clock. Presumably she was looking at it, but there was no way to tell. "A real pendulum in a real clock is not quite so regular as we would wish it. It does not go on cycling to and fro, up and down, without end. If clocks would run forever unwound, why should you have to wind them? A real pendulum is a much more complicated thing than a single atom or than the idealized pendulum that you imagine it to be. It contains many atoms and each of these may use a little of the total energy. Every atom can take some of the energy, and each may use it in a somewhat different way. A real clock will lose energy, or rather it appears to lose it. Actually it goes into the unperceived random motions of individual atoms. The energy becomes *dissipated* by friction at its bearings and even from the resistance of the air through which its pendulum passes. Observe this pendulum carefully."

Scrooge strained his senses and found to his surprise that they seemed to have become vastly more sensitive than he had ever known. He could hear the swish of wind as the pendulum passed through the air. He could hear a squeaking groan from the support and could even sense the tiny rise in temperature as friction heated the bearings of the rocking pendulum.

"All of the effects that you now sense will resist the motion of the pendulum. They all steal energy from it. The amplitude of each successive swing will decrease until the pendulum is finally at rest, hanging motionless below its point of suspension. All the energy that was in the swing will have gone, as it has been converted to other forms. Whatever path it may take, it will eventually end up as heat. The effect is called the increase of entropy," added the Shadow. "Whatever useful or interesting form of energy you may have to begin with, sooner or later it tends to end up as a random background of heat.

"This steady increase of entropy, this random dissipation of energy as heat, happens whenever constraints do not prevent the energy from converting freely between many different forms. When you look at complicated situations with many atoms and few constraints, you find that you can easily distinguish past and future. Consider the falling meteor which you observed earlier. When the meteor struck the earth its kinetic energy became distributed in many different ways: as sound, as light, as the movement of the wind and of the ground. In each form, however, the energy ended finally as heat, and the only overall effect has been to produce a tiny rise in temperature. Might it not happen that this heat should be reconverted to kinetic energy? All of these stages by which the energy was dissipated involved interactions between atoms, and such interactions do not distinguish past from future. If every step of the process is reversible, might not the entire process be reversed? Might this not happen as I shall now show you? Watch and observe!"

The Shadow stopped speaking, and for a moment there was silence. Scrooge watched as he had been instructed but initially could see nothing remarkable. Eventually his attention was caught by a very slight change in a charred branch nearby. Almost imperceptibly, the very tip of the ash-covered wood had developed a small spot that was intensely black. Slowly the black patch grew and spread, and small flames appeared along it. At any rate they flickered and fluttered like flames, but they were dull to look at. Where a normal flame gives off light, these did not. Scrooge realized that they were in fact *drinking in light*. These "anti-flames" were drawing in light from the surroundings and applying the energy involved to the creation of new living wood. The dark flames spread and strengthened. Other branches also burst into black flame, and a faint stirring of the air became apparent. As Scrooge watched, the stir of air strengthened to a wind that rose steadily in intensity, blowing inward toward the position of the meteor strike. In this wind the initial flickerings strengthened into crackling blankets of black which engulfed the branches. Scrooge was intrigued to note that, whereas normal flames are carried outward by the gases escaping from the crackling, popping wood, these dark flames streamed inward and were absorbed by the burning branches, which in the process gradually lost their blackened aspect.

Scrooge once again felt the ground sway beneath him as tremors raced through the earth, moving inward toward the point where the meteor had struck. As the ground tremors surged past he saw fallen trees thrust ponderously from the ground to stand again firmly upright upon their roots. Here and there blazing branches which lay on the ground beside the trees would rise abruptly into the air and attach themselves to trees, which became whole once more.

Clouds of smoke and dust obscured the scene, drifting steadily inward toward the center. A distant echo of thunder could be heard that steadily swelled to a crescendo and culminated in an ear-shattering crash. Shaking his head to clear it, Scrooge looked toward the center of the disturbance and was amazed to see that the smoke and dust had gone from the atmosphere while the trees all stood upright and undamaged. At the very center of the former disturbance he saw the fallen meteor, now fallen no longer but rising swiftly from the ground. It blazed intensely *black*, soaking in vast amounts of heat and radiation from its immediate surroundings as it soared upward. Scrooge watched the dark shape rise into the sky until it was lost in a bank of cloud which dulled briefly as light was absorbed from within by this inversely flaming transit.

"Now that," remarked the dull, empty voice by his side, "is something that you do not see very often. Indeed you do not see it at all. It illustrates how a complete reversal in time might appear on a large scale. You must agree that nothing like that is very likely to happen in reality."

"Unlikely!" retorted Scrooge, "It is totally *impossible*. Nothing like that ever happens."

"No, not impossible; merely *unlikely*," answered the Shadow. "Nothing like that does happen, I grant you but this is not because

it is actually impossible. It is not because there is some basic law of the universe that prevents the transfer of energy from working to undo a catastrophic process in the way you have just seen. It is simply improbable."

"It cannot be merely improbable. Improbable things do happen from time to time, but something like that never does," retorted Scrooge. "I can tell the impossible from the improbable, I assure you!"

"I would not be so sure of that. It depends on just *how* improbable something is. There is improbable and *improbable*, you know. If you buy a raffle ticket it is improbable that you will win first prize, but *someone* will. If you tried to find one particular grain of sand on a beach by picking up a single grain at random it is improbable that you would succeed. In this case it is also improbable that anyone else would succeed either. The reversed sequence I have just shown you is much less probable still. That **it** should happen in reality *anywhere* among the stars of the universe, at *any* time throughout its entire history, is *much more* improbable than the likelihood that you should choose by chance one particular grain of sand from *all* the beaches upon the earth. It is, as I said, not very likely at all."

The shadowy figure by Scrooge's side paused for a moment, overcome by her contemplation of the huge numbers involved. She gave a hollow sigh and continued. "It is *so* improbable that you may be quite confident that it will not happen, but only because it is so *improbable*. The fact that something does not happen is not proof that it *cannot*; only that it is unlikely and so *does not*. What, after all, is to distinguish the impossible from the last extreme limit of the wildly improbable?"

"Well then," pursued Scrooge, "if we grant that energy is allowed to transfer in one direction just as readily as in the other and that such a reversal as you have shown me is in fact *possible*, why should it be so very unlikely? If the basic processes are completely

⟪ SECOND LAW OF THERMODYNAMICS ⟫

The second law of thermodynamics is often stated by saying that disorder will always increase in an isolated system. As a physical law this is generally true, though it is dangerous to use it as a metaphor and to say that everything is doomed to failure.

In statistical mechanics this law depends on the fact that we are totally unaware of much that goes on around us. Everything is made of atoms and molecules, which are in motion. In order to describe matter completely (from a classical, nonquantum viewpoint) we would have to say exactly how *every single atom* is moving. This is technically called a microstate, but in practice we are totally unable to tell one microstate from another. We do not *care* what individual atoms are doing.

If it is easy for energy to transfer quickly from one possible microstate to another, then one is as likely as another. The central notion of equilibrium thermodynamics is that every accessible microstate is equally likely. We are not aware of the individual microstates and see vast groups of them as the same physical system. The most likely system is the one which classes together the largest number of microstates. It is thus the most probable, but the numbers involved are so huge that *most probable* is much the same as *completely certain*.

It may be shown that the number of microstates for hot and cold bodies in contact is much less than for two bodies at the same temperature, so heat will flow from the hotter to the colder. In the process the hotter cools and the colder warms up, so eventually there will be no hot or cold: everything will be at the *same temperature*. This is the **Heat Death**.

reversible, as you tell me they are, I would have thought any happening should be equally likely to go one way as the other. Why then is some state in which the energy is distributed widely as heat so much more likely than the one where it is all tied up in the kinetic energy of a meteor?"

"The one state *is* no more likely than the other," came the dull-voiced reply.

"Humbug!" retorted Scrooge. "You have contradicted yourself. First you tell me that returning all the energy to the flight of the meteor is very unlikely. This I can well believe as I have never observed anything of the sort to occur. Now, however, you are saying that such a return to the initial state is no less likely than a final state in which all the energy is distributed widely as heat. That does not sound at all reasonable to me. It contradicts both my experience and your previous remarks."

"Perhaps that is how it seems, but the statement I made is true. What I have said is that converting to any *particular* final state in which the energy is distributed as heat is no more likely than a return to the original conditions. You should realize, however, that in such a state we must say exactly how *every* atom in the vicinity is moving. Such a state must describe *every* single atom and *exactly* how it moves on a microscopic level. This detail of description is commonly called a *microstate*. Any one such state, so precisely specified in advance, is not at all likely. There are very, very many such states possible, however, and to you they would all look exactly the same. You are never aware of what is going on in this sort of detail or anywhere close to it. There are very, very many indistinguishable states available to any object that that has reached such a haphazard condition.

"However much the internal condition of the object may change from one state to another, though any one state may be just as likely as any other, it is very unlikely that the object should ever happen to change to one of the *comparatively very small* number of states that *you* could distinguish. It is not *impossible* that this should happen, but it is most unlikely because the odds are so strongly against it.

"The condition in which energy is distributed as heat, the random motion of all the atoms, is far and away the *most probable* condition. In practice this means that, sooner or later, all the energy will become distributed as heat; and when it does, by and large it *stays that way*. This is the **Heat Death**. It is the ultimate fate of the universe."

CHAPTER 3

Heat Death

"Heat Death!" Scrooge echoed the words in some dismay. He remembered how he had imagined this to be given as the destination for his last train journey, and now that he heard it again the name seemed still more threatening. "That sounds a fearfully apocalyptic vision of the future. Are you telling me that science is predicting for us a fiery end, like some medieval vision of the flames of Hell?"

"In no way," answered the hollow voice of his companion. "That is not at all my meaning. I am not speaking of the World's death by intense heat but, rather, of the death of heat itself, or at least of any differences in temperature. In the present state of the universe you find much variation between hot and cold, and it is these differences that drive all the variety of motion upon the earth, including life itself. In the final Heat Death these differences will even out, and the Heat Death might better be called the 'Lukewarm Death.'

"The primary source of life and movement upon your earth is the Sun, that ball of intense heat which casts its rays upon the earth. Light from the Sun powers the growth of plants. Animals eat the plants, or in some instances they eat other animals which have eaten the plants. In either case the source of energy which allows them to move and live has come from the Sun. Some energy from the Sun became locked into plants in times past, and these plants have subsequently decayed to give your so-called fossil fuels, coal and oil. Such fuels now provide energy to run your cars and industry and to generate most of the electrical power which you use so freely."

"Our energy does not all come from fossil fuels," protested Scrooge. "A great deal does, I grant you, but some is from solar or hydroelectric generators, some from wind farms, and some comes from nuclear energy."

"Of those you have mentioned, the first three derive also from the Sun. In the case of solar energy this is self-evident. The other two sources are powered by the weather, the movement of wind and water around the earth. This movement also is caused by a difference in temperature: the temperature difference between the day face of the earth, which is being heated by the Sun, and the night face, which is in shadow.

"Your last example, nuclear energy, is indeed different. This does not derive its available energy from the Sun but from energy levels within the nuclei of atoms, where there is an even more extreme concentration of energy than is available in sunlight. The effective temperature in this case is higher yet than it is for the Sun."

The massive figure of the Shadow stood upright in the wan sunlight which illuminated the countryside. The light fell upon her and was totally lost. None of the light escaped, and it did nothing to illuminate her form. With his new knowledge, Scrooge saw in this figure an allegory of the earth as it drank in light and power from the Sun but was always in need of more. His thoughts were echoed by his dim familiar.

"As you can see, energy is steadily flowing from regions of high temperature to cooler regions, so that in the process the hot regions become cooler and the cold regions hotter. Eventually they will all be at the same temperature. All regions will then be in equilibrium, so that there will be no net flow of energy between them. You must expect that, sooner or later, everything will have the same temperature. This is the same as saying that the atoms in every object will, on average, be moving with the same energy."

The shadow moved closer to Scrooge and her voice, usually so dull and unemphatic, seemed almost to take on an lively note. "On the large scale, the macroscopic scale of which you are aware, everything will assume a final, dreadful, total uniformity. You may look upon the future and behold: It will be boring.

"This is not to say that everything will be uniform on the microscopic scale. At that level there will be much the same variety as before. Not all the atoms in any body will have the same energy. They will continue to exchange energy, some gaining and some losing, so that the body will be continually changing from one state of internal motion to some other, but you would be totally unaware of this.

"Blocks of material that are at rest and have the same temperature may seem to you to be totally inactive. They do nothing and do not change, or so it seems. That is only the surface seeming however. The atoms of which they are made are still active; indeed the individual atoms are as active as they are when great changes are visibly in progress. Within, the material atoms will collide with one another. One atom will lose energy, and another will gain some. This goes on continually, with the manner in which energy is shared among the atoms forever changing. There will be just as much energy being transferred between atoms when everything remains at constant temperature as there was when the meteor struck the surface of the earth, but the energy is more spread out and the effects are not apparent to you."

She stopped and peered keenly at Scrooge. Scrooge had never before been stared at keenly by an indecipherable silhouette, and he found the experience somewhat unnerving. "The energy will end up distributed more or less uniformly between the random motions of all the atoms. This does not happen because of any absolute law. It is simply because such an outcome corresponds to so many different but indistinguishable microstates of the system that there is really no chance of finding anything else.

"Any particular microstate for the final system is improbable. If you were to specify in advance exactly the way in which every atom should be moving, then it is *exceedingly* unlikely that your prediction will be correct. It is most unlikely that any particular one out of the vast number of possibilities will actually happen. But *some-*

thing will happen—that you may guarantee. Whatever states do occur will not last for long but will convert to others, because energy
is still being transferred. You cannot predict in any detail exactly
how the atoms will move, but move they will, dividing the energy
between them in some way. The microstates of the world will continue to change as often as ever they did, but to you they will all
look the same. There will be no evident effect from the changes.
That is the important point. You cannot see the individual atoms
and are only aware of their *average* behavior, their *average* energy,
which you call temperature, and this will scarcely change."

"Now wait a minute," protested Scrooge. "You say that energy
will become uniformly distributed because there are many more ways
in which this can happen, but why should that be so? Why is it so
likely that the energy will spread so randomly among the motions
of all these atoms?"

"It is a question of freedom. If an object is moving from place
to place then there is a certain amount of kinetic energy which is
tied up in this obvious and visible motion. As the body is made of
atoms, then some of the kinetic energy which belongs to each of the
atoms will be bound up in this motion."

"That is the point I do not follow!" interrupted Scrooge. He
wanted to make this clear before the Shadow could begin some new
discourse that would leave him little the wiser. "You tell me that
when the energy is lost as heat, or should I say tied up as heat, then
it is distributed among all the atoms in the body. Now you are saying that the kinetic energy of a moving object is also distributed
among the atoms. In what way do the cases differ?"

"As I said, it is a question of freedom. When the atoms share
the kinetic energy of the moving body, they have no freedom in the
way that this energy may be shared. The atoms *must* all move forward with the same average velocity as that of the whole body because each atom is a part of the whole and must not be left behind
or rush in front of the rest. They have no choice. In the case of heat,
the atoms move randomly and independently of one another. There
are many atoms, and each one is effectively free to have as little or
as much energy as it can obtain and to move in any direction it
chooses. The number of possible ways in which any atom may be
moving is large, very large; and if the motions are all independent
then the behavior of one atom does not much limit the choices open
to another. There are a great many atoms in any object that is large
enough for you to see, and the number of permutations of all their
possible states of motion, the possible microstates of the whole ob-

ject, is a large number indeed. The number is inconceivably much larger than the number of possibilities available to atoms which are forced to move in step because they share the kinetic energy of a moving body. As a consequence it will never happen purely by chance that many atoms which have been rushing around at random will suddenly begin to move together in this way. It *could* happen, perhaps, but you can say quite confidently that it *will not*.

"The problem is that there are so many choices open to each atom, so it is most unlikely that it will choose any particular one. It is the problem you would have if you were lost in a maze or labyrinth and moving aimlessly from room to room. The more rooms there are and the more ways of going from one to another, the more difficult it is to find the way out, as you may see."

On these words Scrooge found himself within a square room which contained a table and four chairs and also four pictures, one on each wall. There were four doors. He opened one and found himself immediately outside in the dreary landscape, standing again by the shadow.

"That was hardly very difficult," remarked Scrooge. "The first door I chose brought me out."

"That was because there was only that one room," replied the Shadow. "As I said, it gets more difficult the more rooms there are. See if you find it so easy to get out when there are a hundred rooms."

Scrooge found himself back in what looked like the same room. It had a table and four chairs, what looked like the same four pictures on the wall, and it had four doors. Scrooge chose one at random and walked through it. He found himself in a room with a table and four chairs, four pictures on the wall and four doors. Quickly he chose a door and walked through it. He found himself in a room with a table and four chairs. . . .

Scrooge walked rapidly from room to room. Each door brought him into a room that looked the same as the one before. He tried adopting a plan and always took the first door on the left. He walked on and on until he was sure that he had gone through many more than a hundred rooms, but still there was no sign of escape. He became convinced, though as the rooms were identical it was hard to be sure, that he was moving in circles through the same set of rooms. Accordingly, he began to choose doors at random, though with no more success. He was becoming thoroughly tired of the activity when without warning he found himself again outside with the Spirit.

"There you see," she remarked. "If a maze has a hundred rooms you are not likely to escape quickly. It would take far more than

100 times as long as for a single room, since you may well find that you pass through the same rooms many times as you move around at random. To escape quickly from a maze you must follow some plan, and atoms do not do that. They have no memory, and they never do the equivalent of leaving a trail of bread crumbs to avoid retracing their path. The atoms' movements are based entirely upon chance, and *they* do not have a mere hundred possibilities open to them overall, but millions upon millions upon millions. In practice there is no escape for them."

"You have said that all of these many different thermal states are indistinguishable to us," pursued Scrooge, who had decided to switch the direction of his questions. "How can that be. If there is so much activity and diversity within a body, surely that activity must be obvious when I look at it."

"No," answered his mentor. "You do not really see average behavior at all. You only notice when there are local differences—and that only when they are on a scale that you can see. Your perception of the world is vague and uncertain. If you are in a room that is totally dark, you say 'I can see nothing.' If a light is shone into your eyes you say 'The light is too bright, I can see nothing.' The two cases are surely at opposite extremes of the light you can see, but in each case you can see nothing because there is no contrast. Furthermore, if you take an overall view of something, all the fine detail will average out and you still will not notice it. If you look at a large area of uniform color in a printed picture you may say that you see nothing there, but in fact the colored area is made up of tiny dots of different primary colors, too small for you to see. You just see the overall uniform effect.

"The energy possessed by any object is only obvious if that object has a lot more energy than its surroundings or, contrariwise, if it has a lot less. It is only apparent if there is contrast on a scale that you can detect. It is rather like the material possessions of which you are so proud. Your apartment, your car, every piece of equipment that you own is of the highest quality, and this makes you feel secure in your success. You feel so only because *you* have all these things and others do not. If such material possessions were evenly distributed among the population and everyone had the same as you, they would no longer have value as status symbols and you would have no interest in them. It is because of the differences, because you have them and others do not that they seem significant to you.

"It is much the same with energy. When energy is concentrated in one distinct place or activity and when the average behavior of the atoms in that place is thus different from their average behavior somewhere else, then you have a situation which is striking. It is one that you recognize as being somehow different. The energy which was carried by that falling meteor was afterwards present as the energy of motion of *all* the atoms and molecules which surrounded you. The average energy in one region was the same as in another and so was no longer evident. Come, let me show you."

Scrooge had been looking away, and he turned back at this remark, just in time to see the Shadow bearing down upon him. Before he could react he was completely engulfed by the misty shape. Everything grew dim, and he could no longer see his surroundings. He had a momentary feeling of disorientation, and then his vision cleared. It did not seem to clear completely because his immediate surroundings were still very vague and misty. He looked around him and could see nothing distinct. Then he chanced to look downward, and his vague feeling of disorientation was replaced abruptly by an intense feeling of vertigo. He saw with a sense of horror that far below him were the streets and buildings of a city. His surroundings had seemed fuzzy and indistinct because they *were* fuzzy and indistinct. He was apparently floating within a cloud, far above the surface of the earth.

He clawed frantically at the wispy vapor which surrounded him, trying to find something he could hold and which would prevent him from plunging straightaway to his death far below.

"Relax," came a dull murmur from his side. "Relax, you will not fall. All this is but a vision and you are in no danger." Scrooge looked toward the voice and saw upon a bank of cloud a slightly darker region of mist which was reclining at ease. His companion was still with him and appeared quite relaxed as she reclined on her vaporous couch. "But then," Scrooge thought to himself. "she might well feel at home as she seems scarcely more solid than the clouds anyway."

After this thought had crossed his mind it struck him further that he himself *had not* fallen and was apparently not going to fall, so with increasing confidence he turned his attention to the view below. As he had determined in his first moments of panic, he was floating some considerable distance above a city. He could see the outlines of streets and squares and the line of a river that wound among the buildings. Here and there a church or an office building of more than usual size stood out, but on the whole there was gray uniformity.

Directly below him, now that he could bring himself to look, he saw that there was a large square or plaza. Spread over this was a

pattern of dots in almost continuous motion, such as one might see
when observing the scurrying occupants of an ant hill. He realized
that was approximately what they were, and that he was in fact observing the inhabitants of the city as they rushed about their business. As he watched he was struck by the fact that, although there
was undoubtedly a succession of different people passing through
the square and that some were hurrying to one destination and some
another, yet from his high vantage point the scene looked very much
the same from one time to the next. He knew that each scurrying
dot was a person and that every one was consequently quite different from his companions. The differences that exist between people

are far greater than the differences between atoms, which are in fact very similar to one another, but for all of that the people looked the same from his distant viewpoint.

He watched this view for some time, observing the monotonous sameness which an almost infinite variety will display when it is not clearly seen. People came and went, each one scurrying across the stage of his vision; but apart from a slight fluctuation in the scene, none of this was visible to him.

Suddenly he noticed a distinct change in the scene laid out beneath him. There was some form of steady movement across the square below. He saw a packed mass of dots that moved steadily across the scene as if they were one body which crawled over the city landscape below. He asked himself what could account for this departure from the general anonymity which he had observed and abruptly realized the cause. He was looking down upon a procession. Many people were moving forward with a common direction and purpose, all together and marching in step to the music provided by a band which spearheaded the column. Scrooge realized the point that was being made. The people taking part in the procession were not moving any faster or more energetically than many of the rushing individuals within the anonymous crowd that he had been watching, but he noticed them because they were all moving together, the movement of each being regulated to fit with the other marchers on either side. They were made noticeable, in fact, by their very lack of freedom to move in all directions. He began to appreciate how there could be continuous activity and variety among atoms on the microscopic scale but that nonetheless the whole would appear totally uniform when viewed from a distance.

He reported this new insight to the totally uniform companion by his side who approved his conclusion. "You are certainly right to say that the blur of humanity which you see in the city below is composed of individuals in great variety. Let us descend among them to make clear their differences." As the Shadow spoke, the cloud, in whose support Scrooge had gradually come to rely, was abruptly blown apart by a gust of wind. With his limbs flailing wildly, Scrooge found himself plummeting downward in the long earthward drop which he had previously feared. He looked down upon the distant ground whose features grew steadily larger as they came nearer, but more slowly than he would have imagined. Eventually he realized that the houses and streets below *were* approaching more slowly than seemed reasonable, and that in fact he was settling toward the earth as gently as a falling leaf. When his feet finally touched upon the surface of a city street he was scarcely aware of any jar of impact.

Scrooge wandered here and there along the streets in the company of the Spirit, who now and then would draw his attention to some person as they passed by. They saw an old man, white of hair but with a healthy glowing countenance who came jogging past them with an energy which might put many of his juniors to shame. They saw a young man confined to a wheelchair, his slack face evidence of a sad paralysis. They saw another young man, this time a picture of confident health. He was trying to coax into the back of his car the fabric-covered structure of a hang glider, while already within they could see the heavy air cylinders used for subaqua diving. Later the Spirit indicated a man of middle years who was in the act of stubbing out a cigarette. No sooner had they spied him than he doubled over with an intense fit of coughing, but when he had recovered his breath he took out and lit another cigarette.

Many were the people that they spied who the Shadow selected to call to Scrooge's attention, and many and varied was their form and aspect. "Enough!" called Scrooge. "You have shown me enough. I fully admit that on close examination all the people here are quite different, with all ages, classes, and activities represented. Who could look upon this diversity and know their features and yet say that they were all the same?"

"Look in that window?" commanded the Spirit. Scrooge looked into the glass beside him, but the window was dark and he saw in it nothing but his own reflection.

"I see nothing," he protested, "nothing but my own face."

"You see then the answer to the question you asked," replied the Spirit. "For, you, Scrooge, had knowledge of all these people that I have shown you, and *you* deemed them all the same. You were given the details of each life when they applied to you for insurance, and to your mind they were all the same. Despite their many differences, each was to you nothing more than an Unacceptable Risk. Simple distance is not the only reason for a person to have restricted vision."

Scrooge had to admit that on many occasions he had ignored people's myriad distinctions. Even when he spoke to them face to face he usually concentrated solely on what pertained to his business.

"So," continued the Apparition, "if people, with all their differences, may be seen as but parts of a featureless crowd, an undifferentiated mass of humanity, how much easier is it to miss the merry thermal dance of the atoms. You may fail to see any sign whatever of the varied and ever-changing activity within a solid and think that it is but a dull homogeneous mass, devoid of any interest or activity. Any energy which is within such a body that is in a state of equilibrium will shuttle to and fro forever between the random motions of the different atoms, completely unnoticed and completely trapped. You see only the effects of the energy, which has a low entropy, which is constrained and concentrated to one condition. Such energy can produce visible results. It is the *available energy*, the fraction of the total energy present which is available to make things happen, things which a purposeful and thrusting man like yourself would desire. The available energy is that energy which is still available for you to control and is not yet part of the carefree random thermal dance of the atoms. This 'available energy' may still be bound to serve your purpose and used to make things happen, the sort of large scale, 'everything moving together in step' type of activity which you would class as useful."

"How then comes the difference between these two types of energy?" asked Scrooge. "Why is some available for useful purposes and some not?"

"A good question!" replied the Vision. "Why is any energy sufficiently regulated and constrained that it is available? What is the initial cause of the ordering involved? These are all difficult questions. The distinction is at any rate not absolute, for the Second Law of Thermodynamics tells you that in any attempt to use available energy some portion of it always joins the thermal pool and so increases the entropy, the overall disorganization of the universe. It is a fact sad but true that whenever you use energy in any way to achieve a purpose you will lose some of the energy which had previously been available to you. In any device energy will flow from hot to cold, making both more like to one another."

"But heat does not always go from hot regions to cold regions," protested Scrooge. "I have in my kitchen a refrigerator which will pump heat out of the cold interior into the warmer room."

"In some devices it is true that a little heat may flow in the opposite direction as you describe. In a refrigerator heat is certainly removed from a place that is cold and pumped out to the warmer room, but when this happens a far greater amount of energy must be converted to drive the process and the reduction of entropy in the food that is chilled in the refrigerator is much more than compensated by the increase of entropy when electrical energy is converted to random heat and radiated into the room. It is pointless to

◖◕ Laws of Thermodynamics ◒◗

There are three laws of thermodynamics. The first and second have been mentioned already, the third is not often mentioned and seems to be included largely to make the total up to three. As you will see later Newton had three laws of motion. Three seems to be a popular number for laws.

First law: The energy of an isolated system is constant.

Second law: When an isolated system changes from one state to another, its entropy tends to increase. (Its entropy either increases or, possibly, stays the same.) More energy is lost to random heat and flows from regions of high temperature to cooler ones.

Third law: Changes in entropy tend to zero as the temperature approaches absolute zero. (This last law is rather technical, but it does bring the number of laws up to three.)

leave open the door of a refrigerator in an attempt to cool the room in which it stands, because it will *always* give out more heat than it can extract. If you wish to cool a room you must arrange that the excess heat be dumped somewhere else, well away from the place you are trying to cool. Always and invariably, whenever you operate any sort of machine or engine, energy will be lost to future use. This energy is not destroyed, for energy is conserved and cannot be destroyed, but it will be locked up in the ever-growing thermal pool from which there is no way to extract it.

"Consider this lump of coal," the Shadow said abruptly, pointing to a lump of coal which was lying by the gate of a nearby coal merchant's yard. As she spoke the coal began to glow, and a plume of smoke curled up from it. "The energy which is available to you from the chemical energy bound within the coal could be used for many things. The choice provided by the technology of your society seems boundless. What may be achieved with the energy in a lump of coal is almost magical. Behold the Genie of the Lump!"

The plume of smoke that rose from the glowing coal spread and soared into the sky. It assumed a more solid appearance, at least as substantial-seeming as the Spirit who was Scrooge's companion. Looking upward, Scrooge saw towering over him a gigantic smoky figure, arms folded across a massive if sooty chest. As he looked up in wonder the figure leaned over to address him. "Well, Oh Master, what is your wish? What wonders do you command me to bring forth? A television set perhaps, a CD player or a luxury car? Ask, Oh Master, and your wish shall be granted. But think well, Oh my Master, before you make your choice. Every wish that I grant uses up a portion of my potential, and it can never be replaced."

From the hands of the potent figure flowed a stream of consumer-durables. There were television sets, mobile phones, personal computers: in all a host of such items. Scrooge saw that each object was of the same model as he himself possessed. The pile of desirable items which grew upon the ground before him made an inventory of Scrooge's own possessions. He looked up questioningly toward the mighty form of the Genie and saw to his dismay that the figure was dwindling fast, his massive physique becoming steadily more shrunken and feeble. Before his horrified gaze the figure shriveled and dissipated in the air, and was gone. No more smoke rose, and when he looked into the yard he saw that every piece of coal therein was burnt to a dull, lifeless cinder, emitting only the faintest glow of residual warmth.

"There you see the final problem of technology," remarked the Shadow by Scrooge's side. "You may, by the exercise of sufficient ingenuity, achieve almost anything. However, no matter what you do or how you do it, it will always involve the transfer of energy and also the increase of entropy. Available energy, potentially useful energy, will be lost and converted to inaccessible background heat. Once lost in this way it cannot be recovered. No device, no machine, can operate without the difference in energy concentration which provides useful, available energy. When the available energy has gone, then no machine will run or can run, however ingenious its design may be."

"But surely," protested Scrooge, "even if the supply of fossil fuels such as coal should be exhausted then it will be possible for our civilization to continue in much the same way by using renewable energy sources. In place of coal and oil we can get energy from the wind and the waves and directly from sunlight. We can get fuel by growing new plants, and though we may not be able to wait until they are converted to coal and oil as happened in the distant past we can still burn wood or extract alcohol from plants to give a fuel for cars. I have read of proposals for all these methods of obtaining energy."

"Yes, you might postpone the final loss of available free energy for a long time. You might postpone it for a very long time indeed if you can make efficient use of the radiation that streams from the Sun, but you can do no more than postpone it. The Sun has a vast reservoir of available energy. It is very hot, and energy streams out from it into the surrounding cold of space. It is very large, very hot, and its heat is replenished by nuclear processes within it. It is vast, but it is finite. Like any hot body, sooner or later it will inevitably cool as it loses its energy. In a sense it is cooling all the time because the nuclear reactions which take place are steadily reducing the intense energy that is locked within the nuclei in the Sun. Though it may take a long time, eventually even the energy available from the Sun will all be gone. Look far into the future and see what will befall."

Scrooge looked around him and noticed that the street in which they stood was getting dark, rather rapidly he thought. Scarcely had this thought formed before it was full night, and he saw overhead a sky sprinkled with stars. They were not stationary as stars normally appear, but all wheeled together across the sky with a clear and accelerating motion. As new stars rotated above the horizon they dimmed in a brightening sky, and the Sun burst up in sudden dawn. Hardly a moment seemed to pass before the Sun was high in the sky, with rushing clouds playing tag across its face. But a brief period of light and the Sun tipped below the horizon and the stars again rushed across the sky.

In quick succession day followed, then night, then day. The pace of change ever accelerated until day and night were alternate blinks like a stroboscope in a dance hall. Still the rate of change accelerated until night and day blended into one uniform blur. There was no way now for Scrooge to gauge the passage of time, but he was somehow aware that centuries were passing in the blink of an eye. For an unknown period this state continued and then, without any warning, his sense of time returned to normal and he was standing in broad daylight. The Sun looked just as it always had, but his surroundings had changed. Where before he had been standing in a city street, he now found himself in open country. There was no visible sign of civilization, though on closer examination he saw the eroded remains of masonry protruding from a grassy mound nearby. He had a thought to examine this more closely when, again without warning, he found that he could see nothing but the blur of advancing centuries that he had experienced before.

How much time passed during this remission of its normal rate of passage he could not tell, and he was half in a daze when the world abruptly resumed its usual pace. It was now night and the night still looked much like any other. The stars were in their usual positions as far as he was able to tell, for he had never spent much time in star-gazing. The Moon was high in the sky, and he saw no great change there either. He had time for but a brief look around him before the state of change was upon him again.

This became the pattern of his existence. He spent unknown periods in a timeless state (though somehow he was certain that many millennia passed on each occasion), with each period punctuated by a brief pause in which he had the opportunity to view the current situation on the earth. From one pause to the next he saw great changes in the land about him. The sea, which had been visible on the distant horizon in one glimpse was on the next lapping in a deep bay near his feet. The next time he had the opportunity to look around him in daylight the sea had retreated again to become a distant glimpse of blue, and he stood in the middle of a wide level plane. On the next occasion he saw that a range of hills had risen between him and the distant ocean. These frantic changes in the very structure of the land served to convince him of the immense periods of time which passed between each pause in his flight to the future. He saw nothing in all this to show whether or not humanity still inhabited the earth. Sometimes he glimpsed shapes flying across the sky, but whether they were distant flying machines or some flying creature unknown to him he could not tell.

In all of this frantic change the skies above seemed initially immune, but gradually he became aware of change here also. When he saw Day, then the Sun looked much the same, though if anything a little brighter. When he saw Night, the Moon seemed to have shrunk or grown steadily more remote on each successive view.

After some period, whose vastness he could not begin to appreciate, he could not readily detect the Moon's presence at all. "Whatever has become of it?" he mused aloud. He discovered something else that had not changed when he received an answer to his question. The Shadow, who was after all the source of this experience or illusion, was still by his side.

"The Moon has left the earth," she replied. "Throughout all the long passage of the years the Moon's gravity has reached out to the earth and caused tides within the oceans and even within the solid bodies of the earth and of the Moon itself. The drag of the tides has acted as a brake applied to the earth's rotation and, in the process the lighter body, the Moon, has taken up some of this rotation and in so doing has been flung ever farther outward from the earth. This tidal braking stirs up the waters of the oceans. The end effect is that it heats the water very slightly, just as the brake pads in a motor car will heat when the brakes are applied. This conversion of the energy of earth's rotation is slow, very slow, but it is sure. Slowly and steadily the speed of the earth's rotation has fallen, and the Moon has moved farther away. Not all of the earth's loss of energy has been the Moon's gain. As always, much of the energy has been dissipated as heat, which was produced in the grinding of the tidal motion. Even when the Moon is no longer nearby, the earth still has tides raised by the Sun's gravity, and these too will bleed energy from its rotation until it is locked in position with one face turned toward the Sun, just as the Moon kept one face always toward the earth."

Scrooge made no reply to this information. Somehow there seemed very little to say. Eventually he noted that the Sun was hanging in the sky on every occasion when he could stop to look, and that there was apparently no more night. He guessed that the earth was no longer rotating and that he happened to be on the face which was turned toward the Sun. For many periods the scene was little changed. He had the impression that between each successive view a greater time had passed even than before. Then, in one of the periods of blurred vision which accompanied this passage of inconceivable eons, the diffuse light abruptly became redder and more intense. When again the mad progression paused and he could look around, he saw a vast red orb which seemed to fill half the sky, pouring out torrents of radiation, which had charred the earth around.

"There you see the next stage, which began when the Sun had converted to helium all the hydrogen it contained." With an inevitability which seemed hardly less than that of the Heat Death itself, the Shadow was by his side, providing an explanation of this new sight. "The hydrogen provided the fuel for the nuclear reactions that kept the Sun burning throughout the millions of years that the earth existed, but when it had gone they ceased. The Sun once again collapsed under its own gravity and the release of gravitational potential energy made it hotter than ever before. This new extreme of temperature ignited a fresh nuclear process, and now the helium in turn began to burn with a fresh release of energy. This type of nuclear burning is faster and pours out energy at a furious rate. In the process the outward pressure of the radiation has inflated the Sun to hundreds of times its previous size, as you can see. The supply of helium in its turn is limited, so of course this huge output of energy cannot last for as long as did the previous state."

During his next period of rushing madly toward the future Scrooge saw this intense red glow suddenly flick off and when he next could see around him the landscape was cold and dark. He thought that the earth might have turned so the Sun was no longer visible to him, but when he suggested this his companion corrected him.

"You can still see the Sun. There it is near the horizon to your right." Scrooge looked and saw a small bright speck, little different from the other stars in the sky. "The Sun has used up all the helium fuel and has then passed quickly through the other stages which follow. Now its nuclei have yielded up all the energy which can be extracted from them. Its nuclear fuel is dead, and it has no more sources of available energy remaining to it other than gravity. It can do nothing but cool steadily and collapse, shedding most of its remaining heat until finally it is in equilibrium with surrounding space."

At his next break Scrooge could see no sign whatsoever of the Sun's final remnant. There were still other stars in the sky, but as period followed period these became fewer. From time to time a new star would appear and live for a span, a vast time by any normal reckoning but which now seemed but a moment to Scrooge, the everlasting sightseer. Finally each star would flare up to an intense brilliance and then vanish and with each disappearance the number of stars grew less. The universe became very dark and empty.

"There you see the final equilibrium," came the inevitable voice of the Shadow. "There is the same amount of energy as ever there was, but now it is spread across the empty vastness of space much as the meteor fall that you observed spread its energy throughout

the more local surroundings on earth. As was the case then, the energy is no longer evident to you because it is no longer concentrated and there is no contrast. Everything has reached the same temperature and so it will remain. There is nothing more to say, and the life of the universe is at an. . . ."

The dull voice stopped abruptly. Though tedious and dreary, it had provided the only companionship left in this bleak extremity of existence, and now it too was gone. The night was dark, though Scrooge realized that thought was not strictly correct. Now there was neither day nor night, for that implied contrast, and there was no contrast. There was only darkness and, now that the protection of his mentor was withdrawn, the beginning of a terrible cold. For want of any better plan he cast about him in the darkness and discovered, in the face of all reason, a bed. Without attempting to explain its presence he crawled into it, burrowed beneath the covers, and almost immediately fell asleep.

✿

The Second Visitation

In which Scrooge encounters the ever changing Spirit of Time, who speaks to him not of Being but of Becoming, of change and variation in the Universe. His first visitor had spoken of energy and entropy, of constant quantities and equilibrium. His second visitor talks of inconstancy, of change, and also of creation.

Scrooge is told how motion and even time itself are relative; that time is not absolute but depends upon the motion of the observer. He is told how cause and effect may imply that a complete knowledge of the present will reveal also the past and the future, so that there is then no novelty and no free will. He further learns how even the absolutely determined may still be totally unpredictable and is shown how Chaos may in turn lead to Creation.

The Spirit speaks to Scrooge of Life, the universe, and the mystery of NOW.

CHAPTER 4

Relatively Speaking

When Scrooge awoke he discovered himself to be in his own bed, in his own bedroom. Although he was no longer lost in the intense darkness at the end of time, nevertheless the room around him was still dark. It was so dark that, looking out of bed, he could scarcely distinguish the transparent window from the opaque walls of his chamber. He was straining to pierce this darkness when the chimes of a neighboring church struck the four quarters, and thereafter the hour bell sounded with a deep, dull melancholy note. Lights flashed up in the room upon the instant, and he became aware of a Presence.

He was aware that he was not alone, but for all that he found it difficult to observe who or what it was that had come to him. He was quite sure that there was another being in the room with him, but where? He fancied he saw a figure by the door, but straightway it darted to the window. To the window he turned his gaze, but the figure had gone, flitting over to pause by his electric trouser press. Ever as Scrooge tried to focus upon this new visitation it fled before his gaze, moving and swooping restlessly about the room like a demented humming bird.

Scrooge's baffled eyes turned and twisted in his head as he tried to follow the rapid movements of his as yet unseen guest. Just as he felt he would never again be able to look steadily in one single direction, the Spirit, for so it was, came to rest by Scrooge's bedside, scarcely two feet from his face. This Spirit had the aspect of a young man, or boy even, slim and supple of limb. Scrooge could see that without doubt the figure was now at rest, for did he not keep to a constant position directly in front of Scrooge's eyes? It was strange that, although he did not move one iota from this position, he nevertheless did not *appear* to be at rest. The Spirit stood perfectly motionless, but his hair streamed behind him as if swept back by the wind of some headlong flight. He wore a long cloak which flapped behind him also, carried aloft by this gale, which he seemed to experience even in his present stillness. As Scrooge watched and wondered at this wind of passage where there was no motion to be seen, the movement of hair and cloak abruptly changed direction and now streamed out to the side, as if swept by a different sideways blast, though there was still no other sign of motion. Another few moments and, flip flap, the direction changed again, with hair and cloak swept now to the other side.

"Spirit, be still!" cried Scrooge. "How may I learn from you if you will not stand still? I cannot concentrate if I am to be distracted so. Please be still and say your piece, for time is ever rushing onward."

On hearing these words the spirit laughed, a light infectious laugh, full of liveliness and vitality. "Ho, Man! Would you lecture Me on time and motion? I know that in your advice to those who would run a business you are wont to advocate Time and Motion studies, but what do you really know of Time or of Motion? You might say that I am the Spirit of Time and Motion, the Spirit of Change. Where your previous visitors spoke to you of energy and equilibrium, of the science of Being, I come to show you something of the science of Becoming.

"You asked me to be still," he added. "Do you have any notion what you mean by that? How may you tell whether anything is or is not moving?"

"What do you mean?" protested Scrooge. "If something is moving I will be able to see that it moves. If I cannot see it move it is at rest. Surely that is evident?"

"Distrust whatever is too evident. All is relative! Come! Come with me, for the night is yet as young as I. Come with me and let me show you." The Apparition stretched out a slim young hand. Scrooge took it and straightaway found himself outside his room and rushing over the city beside his companion. Over houses and streets they soared and very soon descended upon the platform of a small suburban railway station. A train was just about to leave, almost empty at this late hour. They landed on the platform close beside a door of the slowly accelerating train.

"Run!" cried the Spirit in Scrooge's ear. "Run, run as quickly as you can, and you may catch the train before it leaves." Scrooge sprinted along the platform, driven by a city-dweller's reflex action when he sees a train about to leave a station and realizes that he might be about to miss something. He did not stop to consider that he had not had any intention of boarding this particular train, and it was no concern of his if it should leave without him.

"Why am I doing this?" he asked himself aloud as he struggled to stay alongside the door while he turned its handle.

"Faster, don't try to talk," answered the phantom. The apparition appeared to be rather older than Scrooge had first thought, a lithe teenager whose long limbs had no difficulty in keeping up with the moving train. Straining every muscle, Scrooge managed to keep pace with the door of the accelerating train as he swung himself inside, closed the door, and leaned panting against it.

The Spirit appeared within the carriage beside him and said kindly, "you may rest a little now. But I would ask you to note that you have been beside this same door all of the time that you were running. Here, you see, it took all the running *you* could do to keep in the same place."

"But I am not in the same place!" gasped Scrooge, "I am in a moving train and so I am by no means keeping in the same place. We are moving all the time as the train moves forward."

"Who is to say that we are moving now?" replied the Spirit. "What is movement if it is not change in position and how do you determine position? The usual method is to say how far one thing is from another, is it not? If you want to talk to someone else about the positions of objects you must agree between you on a way to measure those positions and to say exactly where anything may be found. It is much as you would describe the position of some geographical feature. You would specify its latitude and longitude, which are distances measured along two directions at right angles. You must define a *frame of reference* in order to give the directions you will use. The position might be up a hill, so you would also have to give a distance along a third direction at right angles to either of the other two, an altitude above sea level. In general, if you are to describe something that moves freely in three dimensions, then you must specify the three rulers which you will use to measure its position. Each of these rulers will point in a direction at right angles to the others, and all meet one another at a common *origin,* the point from which you will make the measurements along the three different directions. The rulers, or *axes*, are imaginary, but their point of common origin and the directions along which they lie are real. The three distances from the origin to any point, as measured along these three directions, give the *coordinates* of that point.

"If all your points lie on a flat plane surface, like a map, then you can describe them with only two coordinates. You can imagine the surface to be marked out with a rectangular grid from which you can read these coordinates. This is like the grid references of latitude and longitude for hills or crossroads which you may read from a map. Observe!"

The Spirit waved his hand, and a grid of fine green lines sprang up in the middle of the carriage, like computer graphics superimposed upon a film. In this case the mesh of lines formed a flat sheet which hung in mid-air in the center of the carriage. Scattered here and there over this mesh were a number of bright, stylized trees to indicate that it represented the landscape. A tiny toy train, with a locomotive and two brightly colored carriages, crawled steadily across the grid.

"There you can see that stationary objects, such as trees, are keeping the same position on this grid. The train, on the other hand, moves from one grid position to another because it is *moving* relative to the grid. Let us add another moving object."

Beside one of the carriages appeared a tiny running doll-figure, stumbling and waving its arms in humorous animation as it struggled to keep pace with the train. Scrooge was not at all sure that he found it amusing. "Now you have two objects that are moving, at least as measured on the grid of the landscape. That is not the only grid you could have chosen, however. You could, for example, have fixed your origin to the front of the locomotive and chosen a different grid."

A grid of pale red lines sprang into being. These moved across the green mesh of the landscape so as to keep in step with the train. Since this new grid was fixed to the train, neither the train nor the

☙ FRAMES OF REFERENCE ❧

You may describe the position of an object by giving a coordinate *x*, which is the object's distance from your chosen origin as measured in your chosen *frame of reference*. When the object moves with respect to the frame then this distance will become greater as time goes on (assuming that the object is moving in the appropriate direction) and so

$$x = x_0 + vt$$

where x_0 is the position of the object when you began to time it, *t* is the time that has passed since then, and *v* is its velocity with respect to the frame.

If you look at the object from a frame of reference which is moving with the same speed as the object, then it will appear to be at rest. It speed, *u*, as measured in this new frame will be its speed, *v*, as measured in the previous frame, minus the relative velocity, *V*, of the two frames. A measurement of the velocity *u* will yield the result given by

$$u = v - V$$

Where the second frame is moving with the object, we will have $v = V$, and *u* will be zero; but the relation holds just as well in cases where *v* and *V* are different. It gives the rule for *Galilean* addition of velocities.

---

tiny Scrooge-figure changed their position as measured on it, and all moved forward together.

"There you have a fixed green grid, more usually called a *frame of reference*, which is at rest with respect to the landscape, and a moving red frame of reference, which moves with the train. But who may say for certain that it is the red frame that moves and the green that is at rest? Might not the red frame be stationary?" As he spoke these words the flowing lines of the red mesh halted, and they and the train were abruptly still. Since the red and green lines had a definite *relative* motion, this resulted in the green lines moving in the opposite direction and carrying the token trees backward with the same speed as the train had shown before.

"The train appears to you to be stationary. Might it not actually be so? Can you see the train move? Look around you. The seats, the luggage racks, the door: can you see any of them moving? Look out of the window. You see the houses beside the railway tracks. Don't you see that they are quite clearly moving backward with ever increasing speed. Wouldn't you say that this carriage is at rest, and that it is the world outside that is moving?"

"Of course not!" answered Scrooge. "It is self-evident that the world does not move. I know that!"

"But how do you know? You said before that if something is moving you will see it move. Well here you see that the world outside is moving while you do not see the carriage move at all. Does not that show you the carriage is at rest?"

"No it does not," Scrooge maintained, though it was becoming difficult to argue convincingly. He felt quite sure that the train *was* moving and the houses by the track side were not, but his own earlier statement did appear to contradict this. As he pondered this, the train swept round a bend in the track and he was flung forward into the middle of the carriage. "That is what proves it!" he cried out. "If the carriage were at rest then we would not feel any jolts or disturbances such as that."

"Fairly said!" replied the Spirit. "But even if you do have that reason to say that the carriage is moving, it does not follow that the ground is at rest. Might not *both* carriage and ground be moving, though with different velocities? Whenever there is a *change* in your motion, then you will feel an acceleration and can tell that your motion has changed. You could tell that the train had gone round a curve because you felt the acceleration that accompanied the change in its direction of motion. However, if there is no change in the movement you will feel no effect.

"Motion is relative. If you see that something is always the same distance from you then it is at rest *relative* to you. If you see that the distance is getting less and less as time goes on, then it is moving toward you; or alternatively you may be moving toward it. You can tell that there is relative motion, but you cannot say who is moving and who is at rest. Change in position with time is an indication of movement; indeed it defines movement, and it defines speed. If you were to talk about the speed of this train you would probably describe it in *miles* per *hour*. "If, as now, you see that the rest of the world is rushing past your window, then this tells you that there is relative motion, but that is all it tells you. You may choose to believe that the ground is stationary and the train is moving, but what reason have you for such a belief?"

Scrooge felt that he had such a reason and answered "Does not that sudden lurch the train gave on the corner prove that we are moving? You said yourself that it was evidence of movement."

"It did indeed serve to show that there was a *change* in the movement, but that does not necessarily mean the train is moving *now*. Perhaps it was moving before, but after the change in motion it may now be at rest. Perhaps the train was not moving before, but now it is. Perhaps the train was moving and is still moving, but perhaps the earth is moving also, you cannot tell. You can certainly tell that there is movement, but it is all relative. The train moves *relative* to the ground. It was moving before in a different direction *relative* to the way it is moving now. You can observe relative motion, no question about that; but you can never detect absolute motion. You cannot say what is or is not stationary."

Scrooge became irritated with this pedantic argument. "Surely it is obvious enough? Trains are designed to move, so obviously the train is moving. The earth is very large, and stable, and so obviously the earth is at rest. Is there any need to argue further?"

"You may feel it is obvious that the earth is unmoving, but why is this necessarily so? Is it merely because the earth is so large and extends so far about you that you feel somehow it must be at rest? You know it is difficult to change the motion of anything that is large and heavy, so unconsciously you feel it cannot be moving. Let me show you something that may change your mind. Come with me."

Scrooge was expecting some magical transformation such as he had experienced before, but instead the Phantom advanced toward him. Scrooge was suddenly aware that the Apparition had altered further since he first set eyes upon him. The Spirit seemed to have grown and matured in that short time and was no longer a slight lad, but was now a tall and muscular young man who towered over him. This aggressively physical Ghost seized Scrooge with a powerful grip, flung open the door of the train and pushed him bodily out of it. Fortunately, the train was slowing as it passed through another deserted station, and Scrooge landed staggering upon the platform, stumbling as he tried to rid himself of his forward momentum. He stood upright and turned to the Spirit to protest at this handling of a prominent member of the business community. The altered Vision paid no heed to Scrooge's protests but hustled him unceremoniously from the station.

They emerged onto the parking lot of a large supermarket. The store was deserted, as was to be expected since it was, after all, the middle of the night. Street lights shone over the wall which surrounded the car park and by their light it was possible to see that the great expanse of wet tarmac was almost deserted. There were in all but five vehicles parked there. In the shelter of a far wall three cars and a small van huddled together as if for mutual comfort, trying to escape the attention of the fifth vehicle. This fifth vehicle was parked in the center of the park, though it would be fairer to say that it *completely occupied* the park. Filling almost the entire area was a looming saucer shape. Distant street lamps reflected with an oily sheen from the rain-slicked metal surface of its great domed hull. The vessel rested on massive hydraulic landing jacks, and even in the dim lighting it could be seen that these had been forced deep into the tarmac surface of the parking area, much to the dismay of the supermarket's manager when she discovered the fact on the following day.

As Scrooge stared in wonder at this astounding vessel, a panel opened high in its body and from this a flight of metal steps settled to the ground. Down these came a small figure. Though perhaps, thought Scrooge, the newcomer only appeared small in comparison to his enormous vehicle. The figure reached the surface of the park and walked toward them. As he came nearer Scrooge could see that the figure was indeed short, being only some four feet high. It was also somewhat unusual in being constructed entirely of metal.

"That is our pilot," remarked the Spirit. "He is, as you can see, a robot, but he is very experienced. He has been around." It certainly looked as if he had been around and that for some time. He appeared distinctly worn, his body dented in places and a little rusty around the joints. When he spoke his voice came creakingly, as if from long disuse, and he spoke with a distinct lisp.

"Hello," he said, "nithe to thee you. Would you care to thtep on board, and we can depart thtraight away?"

The small metal figure led the way up the steps, down a metal-lined corridor and into what was obviously a control room. One wall was lined with wide instrument panels, before which stood padded seats for the crew. These all faced what, despite the apparently seamless metal exterior of the craft, looked to be a large window, or at least a view screen of some sort. It showed a dark and rather depressing view of the wet car park.

"Right, all thtrap in," remarked the robot, who appeared to double as captain of the ship. Scrooge and his mentor obediently seated themselves and fastened the wide straps provided, though in the case of the Spirit this was presumably just a polite gesture as he was hardly susceptible to damage.

No sooner were they secured than the view of the car park fell abruptly away. There was little sensation of acceleration, but the city dwindled rapidly beneath them. The car park below shrank to a tiny dark patch, and the houses and streets around it diminished to the tiny buildings of a model railway landscape. Soon individual buildings were no longer evident, and the city lay like a great dark map, bisected by the black winding ribbon of the River Thames. In what seemed no time at all Scrooge could see the distinctive outline of the English coast, becoming ever smaller as they soared into the sky. Then they were far above the clouds, risen so far they could see a distinct curve in the horizon. Soon the earth lay below them like a Christmas tree decoration made of blue glass with swirling streaks of white. Between the white clouds they could see glimpses of land. Of England itself there was no sign because it was, of course, under one of the thickest bands of cloud.

"There is the earth," announced the Ghost rather unnecessarily. "Does it seem to you to be stationary?"

Well it did rather. The visible globe of the earth was certainly impressive. It was beautiful even as it hung in space before them. With all of that, there was no obvious sign that it was moving. Scrooge said as much.

"Have patience!" answered his companion, rather irritably. "You must take a longer view. The pace of change is a matter of viewpoint. To a mayfly a day is a lifetime; to a sequoia tree a century is but a brief episode. I am a Spirit of Time, and so I shall now alter your perception of time so you may see change which would normally be too slow for you to detect."

No sooner had the Apparition so spoken than Scrooge could see that the earth was in fact rotating swiftly about its axis. The lands and seas upon its surface were carried from day to night, from night to day, in a continuous dizzy spin. England and other countries flashed by, like the blurred markings on a spinning top.

"Well," repeated the Spirit, "now does it seem stationary to you?"

"If this is truly how the earth spins," Scrooge began, although he knew that indeed it was so. "If, I say, the earth spins in this dizzy fashion, why do we not feel the same lurching that I felt in the railway carriage when it went round a curve?"

"Oh, you do!" replied the Spirit. "As the earth spins, so your city of London is rounding a curve forever, the curve of the rotating earth. This produces an acceleration, such as you experienced in the train, but now it is a constant acceleration because the earth spins steadily and continuously. The acceleration reduces the effect of gravity slightly, but you are not aware of that because the effect is slight and of course you have not felt the strength of gravity when the earth was *not* spinning. The effect is strongest at the equator because there the ground is farthest from the axis of the earth's spinning and in consequence hastens round most quickly. The Mississippi river flows toward the equator and its mouth is farther from the earth's center than is its source. On this basis you could say that the river flows uphill, and it does so because the increasing acceleration from the earth's rotation thrusts the water out toward the equator."

All at once and without warning the spacecraft leapt away from the turning earth and rose far above the plane in which moved the Sun's attendant planets. Now the outside view showed the Sun as well as the earth and several other planets. Scrooge could see that, as well as spinning on its own axis like a top, the earth was swinging in a wide orbit around the Sun.

"Let me ask you again, is the earth stationary? The answer is obviously no. If the earth is not at rest then what may be: the Sun perhaps? You can see here that the earth and other planets are in continuous orbital motion around the Sun, but is the Sun itself is at rest? If we took a yet wider view we would see that the Sun was in a long slow orbit about the center of our galaxy, the great cluster of stars toward whose rim your solar system resides. Move farther out for a still longer view, and we would see that the galaxy was in motion within a local group of galaxies. There is no apparent end to the sequence."

"Then where is there any true stability?" queried Scrooge. "Where is there a still center of the universe from which all distance and motion might be measured?"

"*Nowhere!* It is nowhere to be found," came his answer. "There is no center. There is no absolute position but only relative distances. Neither is there any motion that is absolute; but again it is only relative. It is all relative. You might as well choose to call yourself the center and to measure all distances and all velocities relative to your own position. It is as good a choice as any, and indeed most people see themselves already as being the center of the universe."

As they talked their craft plunged ever farther from the earth and its solar system. They entered a great cloud of diffuse interstellar gas, with no apparent reduction of their tremendous speed. Obviously the gas was very diffuse, but nonetheless at such speed Scrooge feared for the safety of their vessel. On the view-screen he could see a vague glow of unknown purpose envelope the saucer-shaped craft and guessed this must be some means of protection. As he was regarding this gentle glow he was startled by a vivid red flash as a beam of intense light blazed across their path. He was able to see it as a thin, brilliant red line which crossed his field of view because the surrounding gas scattered some tiny fraction of the intense light into his vision. He was about to make some comment when another beam as brilliant as the first lit up the viewport.

"We have company," remarked the Spirit. "There are not one but *two* other ships out there, and we are observing their conversation with one another. They are communicating by laser beams, and you have just seen a preliminary exchange of formal pleasantries. You noted of course that the two pulses of light came from opposite directions."

"How could I possibly note that?" asked Scrooge petulantly. "All I see are bright lines drawn across the dark. I cannot tell which way they are going."

"Of course, of course," sighed the Apparition. "I was forgetting your limited ability to see movement when it is right in front of your eyes. I shall now have to speed up your time sense so that you are more aware of fast movements." No sooner had he spoken than the world around Scrooge seemed to freeze into immobility. The twinkling lights on the control panels paused in whatever pattern they had when the Spirit spoke. Their pilot sat like a metal statue at his controls. Outside the view was similarly frozen also, though this did not make much apparent difference. Such is the immensity of space that there had previously been no changes perceptible to Scrooge's normal time sense.

After a period of total inactivity, Scrooge saw another laser pulse flash across his field of view. Even to his altered time sense the light moved very quickly, but now he could at least observe its motion as gas atoms glowed fleetingly from its passage. The pulse moved from left to right, and he could see the complex sequence of short pulses carrying a message from one spaceship to the other. It flashed past his eyes like an exotic string of brilliant red beads. Shortly after another similar string came back along the same path, but in the opposite direction.

"Have you noted the velocities of the two laser pulses?" he was asked. Of all that surrounded Scrooge, the Spirit alone appeared to move at a normal rate. This was hardly surprising as it was he who had been the source of Scrooge's own altered perception and seemed to have control of time. He appeared now as a more mature and impressive figure, a man past the first bloom of youth, with bushy hair and a drooping moustache, a man of great understanding.

"Yes," replied Scrooge. He had observed the laser pulses rather carefully, since they had been the only signs of movement that he could now detect. "As far as I could tell, each moved just as quickly as the other, though in opposite directions."

"Exactly!" cried the Spirit; and upon the instant the lights on the control panels began again their rapid flickering. Their pilot broke his frozen immobility and leaned sideways in his chair to make some small adjustment to a control. The demonstration was apparently over, though Scrooge was at a loss to discover what, if anything, he had learned.

"Does it not seem strange to you that the two pulses of light travel at the same speed?" pursued his mentor. "After all, one was emitted by a ship which is racing toward us, the other from a craft which follows in our wake. Our relative speeds are very different in the two cases. The ship which is moving toward us has a much greater relative velocity than does the ship which follows in our wake. 'A stern chase is a long chase,' to quote an old nautical saying. If the relative velocities of our vessels are so very different, does it not seem strange that the light pulses emitted by each ship, and moving at the same speed *relative* to that ship, should be seen by us to have the same velocity?"

Scrooge opened his mouth to reply, but the Spirit carried on without pause. "Of course it does. It does seem strange. It *is* strange; strange but true. It is an observable fact, and one demonstrated time and again that the speed of light is always the same no matter who may measure it. No matter how you are moving, no matter how fast the source of the light is moving relative to you, you will always see that light travels with exactly the same speed."

"But that does not make sense!" protested Scrooge.

"It does not have to 'make sense.' It is *truth*, and the truth about the universe need not oblige you by 'making sense.' That is how it is. You may not understand why it *should* be so, but you can usefully consider what are the consequences of the fact. "The fact is that, no matter how you move; no matter what the *frame of reference* from which you view it, light will always be seen to travel with exactly the same speed. That is one observation which is not relative. The speed of light is *special*, and as such it is basis of Einstein's 'Special Theory of Relativity.'

"The special behavior of the speed of light has nothing to do with light as such. It is not the *light* but the *speed* that is special. There is a limiting velocity through space, and nothing can travel faster than this, try as it will. Speed, relative speed, cannot increase indefinitely. If you try to move ever faster by continuous accelerating in a spaceship, then any outside observer would see that your speed steadily converged on this limiting velocity but never exceeded it. No such observer will ever see you traveling any faster, no matter how they themselves may be moving."

"Why need you worry so about outside observers?" argued Scrooge. "How great will you see your own speed to be?"

"You will not see it at all," replied the Apparition patiently. "You can never see your own speed. You can only observe motion of other things as they move by you, All movement is relative, and obviously you cannot move relative to yourself. If everything is moving past you with the same velocity, then you may interpret this as meaning that you are moving and everything else is at rest, but that is an *interpretation*. You can never see yourself as moving either toward or away from yourself. As far as observing relative motion is concerned, you yourself will always be at the still center of your own universe."

"Well then," pursued Scrooge, attempting a different line of attack, "after all that you have told me about relative velocity, I do not see how it is possible for the speed of anything to be the same for all observers. You have stressed to me that speed is the distance covered in a given time, but if someone moves toward me then their distance from me becomes less; if they move away the distance becomes greater. In such case how is it possible that *any speed* may be seen to be the same by everyone?"

ᛞ CONSTANT LIGHT SPEED ᛩ

It was found from experiment that the speed of light is the same no matter who measures it or how they may be moving. This was obviously ridiculous! The *same* speed could be calculated from Maxwell's equations for waves in electromagnetic fields. Again the speed did not seem to depend on how the observer was moving. It is now very well established that the speed of light is constant no matter who sees it and no matter how they may be moving.

Einstein's *Theory of Special Relativity* is based on this experimental fact. The complication comes because it is obviously true that if you move toward or away from something then your distance from it must change. If the speed of light is unchanged, then it follows that, since speed is distance divided by time, so *time* must change also. The consequence that periods of time as well as distances depend on the motion of the observer is the striking feature of *Special Relativity*.

Einstein's theory is more confusing than the earlier Galilean relativity, in which space was relative but time was not. It is, however, the simplest way of making the speed of light constant—and constant it does seem to be.

"Your question provides its own answer. If the observers see different distances but the same speed, and if speed is distance divided by time, then it follows that they must see different times. That is the only possibility. That is the conclusion of the special theory of relativity. *Time is relative.*

"Such is the essence of how light comes to have a constant speed. Different observers do indeed see the light as traveling different distances. If the observer is moving to meet the light then it does not have so far to travel before they meet. If he is running away from it, then it travels farther. However, the time which it takes for the light to travel from one point to another is different as seen by the two observers, and the ratio of distance divided by time, which is speed, is the same for both."

"That cannot be true," protested Scrooge. "Time is time! It is the one thing that we can do nothing about. Time passes at a constant rate and is the same for everyone. Time is surely an absolute if anything is!"

"I shall ignore your remark about the rate at which time passes, since a rate is a measure of how something changes with time. How can you measure the way that time itself changes with time?" replied the Spirit severely. His pedantic tone would have seemed out of place for a vision who had seemed a relatively young man, little older that Scrooge himself; but now he seemed to have aged still further and appeared to be a man of mature years, with a graying moustache and wild, unkempt hair. When he spoke his words carried great authority.

"There is *no* absolute time, any more than there is absolute position or absolute speed. They, all of them, depend on the vantage point from which you observe them. It has not been easy for the universe to make the speed of light constant for all observers. It has had to produce what will seem to you to be severe contortions of space and time in order to do so. The constancy of the speed of light is apparently something about which the universe feels quite strongly."

The Spirit paused and glared at Scrooge, anxious to impress upon him the astounding nature of the fact that light should have this same constant speed for all observers.

"Light is a form of electrical wave motion and as such would seem akin to sound waves in air or ripples upon the surface of a pond. In these other cases the speed with which the wave travels depends on many things: the pressure and temperature of the air, whether there is oil on the surface of the water, all sorts of conditions. This is not the case for light. The velocity of light may be calculated from Maxwell's Equations, which describe all observable electrical effects. The result is a unique value, with no *provisos or quid pro quos*—with no reservations about the motion of the source or of the observer or anything else. This suggested that, despite all that had been believed about the relativity of space, there must, needs after all, be some absolute frame of reference in which light had this unique speed. Nature has avoided the need for any such favored frame by arranging that light has the same speed in *every* frame. The speed of light *is* the same for all observers, and this requires contortions of space and time, which are there to be seen and *are* seen.

◖ THE LORENTZ TRANSFORMATION ◗

The Galilean transformation $x' = x + vt$ allows you to calculate the coordinate x as measured in one frame from the coordinate x' and time t in another frame when two frames are moving with a relative speed v. There is another equation in the set of Gaililean relations which is not usually quoted but is simply assumed, and that is $t' = t$. It is assumed that time is always the same.

The Gaililean transformation will never give the same speed for anything if you measure the speeds from different moving frames. If the speed of light, c, is always to be the same, then you must use a different form of transformation, the *Lorenz transformations*.

$$x' = \gamma(x + vt)$$
$$t' = \gamma(t + \frac{vx}{c^2})$$

Here γ is a fairly complicated factor that is very close to one for low velocities but becomes large as the velocity approaches the speed of light, c. For low speeds, where v is very much less than c, the Lorentz transformations are virtually the same as the Gaililean ones, but at high speeds they become very different.

The various predictions that follow from the Lorentz transformations may all be tested experimentally and, however suprising, are found to be correct.

"This variability of time brings other strange distortions when high speeds are involved. Anything which is moving fast enough for the effects to be obvious will be observed to be shorter than it would be if it were not moving relative to the observer. Though I use the word *observed*, you should note that this is no sort of optical illusion. If you were to look at an object which is moving at a speed close to the speed of light, then it would look strange because it would move a long way in the time it took light from it to reach your eye. You would always be seeing where it *was* and not where it *is*, even as measured in your own stationary frame. After you have made allowance for all such effects which may be put down to the time it has taken the light to reach you, you will still find that the object does seem to be shorter. However you may attempt to measure it, it will turn out to be shorter. As far as you can possibly discover, it *is* shorter. For you the reduced length will be genuine, but any observer who is traveling along with the thing will see it has its normal length."

"How can that be?" protested Scrooge weakly. He felt that he should make some protest in the name of common sense and normality, though his feeling for both these concepts was fast being eroded.

"How can it be?" echoed his instructor impatiently. "You might as well say 'how can it *not* be.' You must realize that Nature is as she is, and your wishes and preconceptions in the matter have very little weight. You cannot usefully argue how or why it might be true or whether it ought to be true. Observation serves to show that the length contraction is true, so there the matter must end."

The Spirit glared at him for a moment and then continued, "More interesting and stranger yet is the effect of *time-dilation*. When you see something which is moving at a speed close to light speed, then you will see time slow down. Clocks will run more slowly, people will age more slowly, everything will take longer to happen. I should stress that this is not due to any change in the things themselves. An observer who is traveling along with them would see no change because there is none. What is changing is the nature of time itself when seen from a rapidly moving viewpoint. Again it is no illusion. Come and see!"

The Spirit spoke briefly to their pilot, who made some small alterations to the controls. Time passed. It did not seem to be very long, but Scrooge had become cautious about judging the passage of time when in the presence of his mentor. Sufficient to say that after some time he saw the growing disk of a distant planet and in front of it a complex metallic shape decked with bright navigation beacons. This grew and resolved in his vision until it could be clearly seen as a large, though irregularly shaped, space station in an orbit around the planet. As they came nearer, the overall impression of random complexity resolved into a host of individual features. There were observation domes and entry ports, bright warning lights, and unintelligible markings on the surface. The whole vast construction was a jigsaw collection of pressurized sections and linking walkways, of piping and external instruments of no recognizable purpose. At one end of the station's central body was an enormous open portal, and entering this was a small spaceship, dwarfed by the surrounding bulk of the satellite.

CHAPTER 5

The Light Barrier

Scrooge's vessel followed the other through the open door and into a large docking bay beyond. Scrooge and the Spirit quickly followed the sole person to leave the other ship. Scrooge fancied that the crew members on duty around the hangar were startled and bemused to see the ship which they had been expecting followed so closely by this strange saucer-shaped craft,  but no one commented or took any action. It seemed that his phantom companion must have confused their minds so that they could not accept the reality of what they saw.

Their quarry was a young man, clean-cut and determined, who looked like a hero from the sort of adventure story that Scrooge had read as a boy. They followed his athletic stride down long metal corridors until he stopped and knocked at a door. After a moment's pause he went in, and they followed. Inside he stood to attention in front of a desk behind which sat an elderly man with a square-cut beard that was almost pure white. The younger man began to make his report in some language totally unknown to Scrooge.

Scrooge looked questioningly at the Spirit standing motionless at his side. It was obvious that the present scene was supposed to emphasize some point, but he could not imagine what it might be. "Well, Spirit, enlighten me if you will. What should I find so remarkable in seeing these two men talk?"

"Only that they are twin brothers."

"Twins? How can that be? One of them is, by his looks, many years older that the other."

"You are right; indeed he is."

"Do you mean then that, although they are genetically twins, one of them managed by some strange freak of medicine to be born much later than his brother?"

"No, of course not. They are twins and so were born at the same time and in the same place. When they had grown, however, they chose different careers. One became a pilot and spent most of his life flying across space at great speeds—speeds which were very close to the speed of light itself. His brother took an administrative post and has remained for most of his life in this slowly moving space station in its orbit about his home planet. From his dull and monotonous post he seldom sees his brother and then only when he is departing, beginning his long acceleration to a large fraction of light speed or perhaps for a short period after he has returned from a previous journey. In the time between these infrequent meetings the 'younger' brother is always on the move and moving at speeds quite close to that of light. Sometimes he is moving quickly away, sometimes returning quickly, and at other times between he will be moving in other directions as he plies his trade between the planets. Always the younger brother is moving relative to his 'elder' twin and always at a speed quite close to that of light. Because of this he ages the more slowly."

"Wait a minute!" interrupted Scrooge. "Have you not been telling me again and again that all motion is relative, and that any one frame of reference may just as well be considered to be moving as any other? In that case should not the space-faring brother have seen his desk-bound twin as moving equally rapidly away from *him*, and so *he* should see his twin as remaining the younger. As you have been at such pains to tell me, the situation is symmetrical, and there should be no reason to choose one frame as against the other."

"What you say would be perfectly valid if we had but two moving reference frames involved," conceded the Phantom. "In that case you would indeed expect a complete symmetry between the two; and because in this case we see a clear asymmetry between the two brothers, a situation such as this is known as the *twin paradox*. It should be clear, however, that such a symmetry need not exist in this case because there are of necessity more than just two frames involved. The results of special relativity hold only for observations that are made from so-called *inertial frames*. These are frames of reference which move with a steady, constant velocity and do not accelerate. You observed in the railway carriage how you were able to detect acceleration only when the train altered its course by going round a curve and throwing you to one side. This distinguished the 'moving' train from the 'stationary' ground or, more accurately, the 'accelerating' train from the 'inertial' ground, since you have no cause to say that the ground need be stationary. The *general theory of relativity* can deal with accelerating frames. It says that you cannot distinguish whether you are in a frame which is accelerating or in an inertial frame in which you feel a *force*. The theory of special relativity, however, deals with frames that are *not* accelerating, and it is by applying this theory that we argue the moving twin ages more slowly."

"How do you argue that?" asked Scrooge. "For example, my cousin is of roughly my age. He was born in the same week as I was, and he lives a dull and uneventful life, with far less activity than mine. I note that as far as one can tell he is still the same age as I. All of my activity has not made me one day the younger."

"You would hardly expect that it should," countered the Spirit. "You must remember that all the strange transformations of special relativity are all because the speed of light is constant. They have significant effects only when the observers move with relative speeds which are very close to that of light. I do not imagine that in all your activity you move quite so quickly as that. If you did you would reach your Moon in just over a second, as those left on earth would measure the time. At such a speed you yourself would register a time that was shorter yet."

Scrooge was temporarily silenced by this thought, and the Apparition continued to speak about the twin who was moving rapidly.

"If two people are moving quickly but steadily relative to one another, then the theory of special relativity maintains that each would see the other age more slowly. If they continue to move apart with this constant steady speed, then the situation would indeed be symmetrical, but if that were the case they would never get together again so that they could make a direct comparison. If they do meet up later it can only be because at least one of them has *not* been moving steadily but has been changing his direction of motion. At the very least he must have turned around to come back. The twin who stayed at home has been sitting placidly in the same inertial frame day after day. He has not been accelerating, and for him the relations of special relativity are valid when applied to each and every one of the different frames in which his brother may be found at different times. He is always looking from the same steady frame, and so he sees things from one single viewpoint. He sees his twin age slowly while he is traveling, no matter what the direction; and when they get together again he finds that his twin *is* much the younger. His twin is forever changing from one reference frame to the other, and it is much more difficult to infer the total elapsed time he would expect to see, but we can be sure that when he returns to his brother's side and is again at rest in his brother's reference frame, in the same time and same place, then the elapsed time as measured in his brother's frame will be correct.

"For the two twins time is indeed relative. Though in practice," he added, "it is not at all easy to accelerate to speeds that are sufficiently great for this *time dilation* effect to be obvious. Come along and I shall demonstrate." The Spirit turned on his heel and left the room, striding rapidly down the long corridors by which they had come. Scrooge hurried along in his wake. The twins did not mark their going any more than they marked their earlier arrival.

The Specter continued speaking as they hastened down the metal-walled passages which led back toward their vessel. "When you consider the singular behavior shown by the speed of light, have you asked yourself what you would observe if you were in a spaceship which continued to accelerate up to and beyond that speed?"

"I cannot imagine," answered Scrooge, "but if the speed of light is so great, then I do not imagine that it can be easy to exceed it."

"Oh, you are correct about that. It is not easy. In fact it is not *possible*. No matter for how long a rocketship, or anything else, may continue to accelerate, it can never exceed the speed of light. This is true whatever the frame of reference from which the ship is observed. The speed of light is the same for all observers, so any speed

⸿ Time Dilation Effect ⸾

The *time dilation effect*, which is the basis of the *twin paradox*, requires considerable care to work out from both viewpoints, but it is quite possible to do so. It has also been tested experimentally and is confirmed to great precision. It has been measured for "atomic clocks" carried round the Earth by fast airplanes. Even the fastest jet plane can fly at only one hundred thousandth of the speed of light, and the time dilation produced is minute. The effect can be measured because the clock is extremely accurate, but it is a difficult measurement. Much more convincing results are obtained by using particles instead of planes. Charged particles such as protons may be accelerated to speeds very close to that of light, and for them time may be altered by a factor of a hundred or more. The factor may be confirmed, since some particles carry a sort of "clock." They suffer from a form of radioactive decay with a very well-defined lifetime. When the particles are coasting round and round at high speed in a circular accelerator, the measured lifetime increases. It increases by exactly the factor predicted by Einstein's theory of special relativity for the known speed of the particles.

You might object that the best tests are only for tiny particles, but what are large objects other than a lot of particles moving together?

less than the speed of light will be *less* for all observers, if not necessarily by the same amount for each one."

They reached the last corridor, the one that led directly to the docking bay, and turned into it. As they hurried down this last stretch, Scrooge made an attempt to understand this conundrum of light speed. "Do you mean that when the speed of a vessel comes close to this extraordinary speed then for some reason its crew would find that it was no longer able to accelerate? Would they discover that its engines would suddenly cease to work?"

"No, of course not! The special characteristic of the speed of light is the way that it is seen by observers in different states of motion, but the crew of the spaceship would not see the ship or themselves as moving at all, because they would *not* be moving relative to themselves. They would always see themselves as stationary, and so they can observe nothing out of the ordinary. The engines would continue to operate as usual. The ship would accelerate, and the crew would feel the acceleration, exactly as they did when the ship was just beginning its period of acceleration, when outside observers would say it was still moving slowly. To those inside the ship nothing would or could have changed. The acceleration would continue just as before."

They reached the end of the corridor and began to cross the hangar deck to reach their saucer-shaped craft. The crew members of the space station who were on duty in the docking bay ignored them, just as they ignored the flying saucer parked in their midst. Whenever their business took them across the area where it sat they would make a wide detour around it, while apparently quite unaware that they had done so.

"Do you mean that the people in the ship would continue to detect acceleration, but that those outside would not? Is the thrust which drives the ship forward not observed from the other frames?" suggested Scrooge rather desperately as they reached their vehicle and began to climb its steps.

"Again no! The thrust of the motors would be quite apparent in all frames. The rocket would still be driven powerfully forward as it approached the speed of light, and all observers would see that this was so. Where you are wrong is in your belief that, just because something is continuously pushed forward by a force, its *speed* must go on increasing steadily!"

Scrooge paused in disbelief, just as he was about to enter the vessel. "How foolish of me!" he said sarcastically. "For certainly I was of the opinion that, if something is propelled powerfully forward, it will probably move rather more quickly!"

His instructor passed Scrooge in the doorway and led him into the saucer's control room. "The effect of a force," he continued calmly, "is to increase *momentum*. My elder sister, the *Mistress of the World*, has spoken to you about momentum and how it appears at low speeds. She told you that momentum is proportional to speed; and as a consequence, when the momentum is increased by a force, so equally is the velocity. At low speed the increase in momentum produced by a force will be seen to be the same as acceleration: an increase in the *speed* of whatever it is propelling. As momentum is proportional to speed, either result is equally true. When a force acts on something which is traveling with a velocity close to the speed of light, then the force will still cause the momentum to increase, just as it would at lower speed. The difference is that, although the *momentum* can continue to increase indefinitely, the *speed* will never rise above the limiting velocity of light."

They had by now seated themselves in the chairs by the control panel. Their metallic helmsman pressed a selection of buttons, and they could hear distant klaxons sounding outside the ship. Those workers who had been in the hangar area fled precipitously. They evacuated the area before it was truly evacuated: that is to say, emptied of air as it must be preparatory to opening the door that led out to space. As the great door slowly slid open, their craft lifted silently and glided out and away from the satellite. On the view-screen they watched the space station's many lighted windows shrink rapidly behind them.

"It turns out," continued the Spirit as if there had been no interruption in his remarks, "that, although momentum is simply proportional to velocity at low speeds, this is no longer true at high speed. The momentum of any object will continue to rise without limit as its speed approaches that of light. If you were to insist that momentum is given by the relation that holds for low speeds, namely the speed of a body multiplied by its mass, then near the speed of light it would seem as if that mass increases rapidly. This is not a helpful view, and it is better to say that the mass of an object is a definite property of that object and has the value which you would measure in the object's own frame, in which it is of necessity at rest. Momentum has a more complicated dependence on velocity than had been apparent before, though this dependence is indistinguishable from that given by the object's mass multiplied by the velocity when that velocity is low."

"But how can the nature of momentum change?" interrupted Scrooge. "If the momentum of a moving body was found earlier to be given by its mass multiplied by its velocity, and if that was right and satisfactory, then surely that is how it is. How can you now say that it is something different?"

"You may never be completely sure that any physical result or equation is totally, finally correct," asserted the Phantom. "Nothing is given to you as being perfect received wisdom. You have to discover everything by experiment and check by repeated observation. You can never prove absolutely that your theories are correct, though you may readily show that they are wrong when their predictions do not agree with what you see. The best you can ever do is to show that your theories agree with everything that you have observed *so far*. It may be that all of your observations have been made in restricted circumstances, and that in those circumstances your theory is a *good approximation* to the true situation. As long as all observations were made at speeds that were very low compared to that of light, the old relation for momentum gave wonderful agreement with observation, but when you move to very high speeds it fails and the need for a new form is evident. The new form must agree sufficiently well with the old wherever that is known to work, but it may predict quite different behavior elsewhere."

"So I can expect no statement of Science to be final?" said Scrooge regretfully.

"I fear not. Any result, however well it may seem to be established, may be hiding a deeper and more remarkable truth within it. This all serves to make the subject more interesting. You find that, rather than a body's momentum remaining strictly proportional to its speed, it can rise without limit even though its speed is limited. Its energy shows a similar behavior. No matter how much energy you give to anything, you can never make it exceed the speed of light. The energy of any object that has mass will rise steadily as it closes in on the speed of light."

"So I can expect all the effects of this special relativity to be insignificant unless I am looking at something which is traveling close to the speed of light," asserted Scrooge who felt that he had at least grasped this point.

"On the whole that is so," affirmed the Spirit. "There is, however, one significant difference between the behavior of energy and momentum, and that is seen, rather surprisingly, at *low* velocity. As any object's speed approaches that of light, both its momentum and energy will rise without limit. As its speed drops to zero, however, its momentum also will fall toward zero, but its energy will not! The formula which describes correctly how energy changes with speed when velocity is high does not give a value of zero when the body is at rest, as you might expect. The energy still has a very large value, that is called the *rest mass energy* of the object. There is a large amount of energy somehow locked up in the mass of any body. One of the conclusions of the special theory of relativity is that mass is energy, and energy is mass. They are in some way aspects of the same thing. Usually the existence of this great amount of energy which is concentrated in the rest mass of an object does not matter because of course you do not usually find that the mass of things changes, and so the energy is not *available*. There is never any obvious change in the mass of objects in most circumstances, except when bits fall off or extra bits are fastened on, and then the change in rest mass energy is all in the masses of these bits. It is not released in any other form.

"If anything that possesses mass should reach the speed of light, then its energy would become infinite, however small its mass might be. There is not enough energy available in the whole universe to allow this. Our next destination should illustrate the difficulties this can cause."

⫷ℭ RELATIVISTIC ENERGY AND MOMENTUM ℭ⫸

When you describe objects that are moving with high speed, it is necessary to use more complicated expressions for momentum and energy.

For momentum p you must use $p = \gamma\, mv$ in place of $p = mv$.

For energy E you must use $E = \gamma\, mc^2$.

The symbol γ above represents a fairly complex expression that increases rapidly as the velocity v becomes close to the velocity of light, c.

The expression for momentum on the left reduces to the more familiar expression on the right at speeds so low that γ is indistinguishable from unity.

The expression for energy reduces to $E = mc^2$ at zero velocity; but when the velocity v is low, but not zero, the energy can be shown to reduce to $E = mc^2 + \frac{1}{2}\, mv^2$

The second term is the normal expression for kinetic energy at low speed.

The first term is called the *rest mass energy* and is much the larger when v is small.

Even when something is stationary it has enormous energy, which is given by the famous equation $E = mc^2$. This was not evident in traditional Newtonian mechanics because the mass of objects did not change, so the energy was not released.

Their saucer sped on through the unchanging vista of space until at last they saw, straight ahead of them, the bright speck of a distant star. This grew steadily more intense until a sun shone in their view screen. The screen showed also that the nearby volume of space was abnormally crowded with asteroids of all sizes, ranging from smallish rocks up to veritable planets. All were concentrated remarkably close together as if they had at some time in the past been collected for convenience. Many of them were very rough looking, all hideously marred by great scars and craters on their surfaces. In some cases the pits were gaping wounds, reaching far into the interior of the bodies as if they had been sliced open to tear out the very hearts of the planets.

Among this clutter of rough astronomical debris there floated one body whose surface did appear to be smooth, although it was not a sphere. It was long and comparatively thin, a swollen needle that drifted amid the remains of the system's planets and from a distance looked more like a rocketship than anything else. It was obvious to Scrooge that nonetheless it *must* be something else because when a nearby planet passed in front of the slender outline it could be seen that its length was fully as great as the diameter of the planet.

"Why are we here and what is that long shape?" demanded Scrooge, who had become rather tired of being dragged, willy-nilly, across the universe at the whim of this fickle Phantom.

"It is a rocketship."

"That is absurd! How can it be? It is enormous. Why, it must be over ten thousand miles in length! However many people is it meant to carry? It could surely transport the entire population of their home planet. Is that what it is designed for?"

"No, it will have but one pilot. The beings who have constructed it have in the process stripped bare the resources of their solar system to build this one tremendous craft, which is to carry a single representative member of their race and boost his velocity as close to the speed of light as they can manage. To achieve this they need to build the largest vessel they possibly can.

"A rocketship must conserve momentum, as must anything else. A rocket takes advantage of this fact to propel itself in empty space,

where there is nothing to push against. You have already been told that when you run upon the surface of the earth, the system of *earth plus you* must conserve its momentum. As you run forward, the earth itself will absorb an equal momentum in the opposite direction, though such is the huge mass of the earth that this backward motion is quite unnoticed. Similarly, the system of *rocket plus exhaust gases* must conserve momentum. The center of mass for the whole system cannot move; but as the exhaust rushes out from the rear of the rocket, so the rocket receives a little forward momentum to balance that carried away by its exhaust. If the exhaust moves *very* quickly, then the rocket will move forward quite smartly, even though it is comparatively heavy.

"If a rocket is eventually to move close to the speed of light, it must start with an initial mass which is very large indeed. This mass must be large enough to carry away an amount of momentum sufficient to balance the immense forward momentum of the final section of the ship. The designers have tried to reduce this momentum as far as they can by abandoning all fuel tanks and other unwanted structures as soon as they have served their purpose, so as not to waste fuel by accelerating unnecessary mass. The ship must have an enormous amount of available energy stored in its fuel and use it as required to provide the high momentum with which the ejected fuel must leave the rocket. All of these considerations make for an initial structure of extremely large size."

"But the whole idea is ludicrous! Why should they ever do such a thing? Why go to such suicidal effort just to approach the speed of light? And with only one pilot!"

"The philosophy of these beings would indeed seem strange to you, Scrooge. It may best be understood by considering their ancient proverb:

' ♌❋✈♎☺ ‼ '

"Unfortunately, this pithy saying is almost impossible to translate. The best approximation I can manage in your language is

"Because it's there!"

They moved unhesitatingly toward the giant ship and turned toward its forward end. Their saucer, which had previously seemed a large vessel in its own right, was now but a tiny gnat flying along the massive flank of the tremendous ship. They sped above the curving landscape of a booster stage which was the size of a continent

and on past further stages of successively decreasing size. Eventually they reached the final booster, which was only about the size of an aircraft carrier, and came finally to the ship itself. This was a tiny speck a mere hundred meters long, perched like a radiator mascot at the front of this tremendous reservoir of energy.

As they came abreast of the tiny pilot capsule situated at the extreme tip of the monster, Scrooge's attendant Specter reached across and seized him by the arm. He experienced a moment of confusion as his view blurred and shifted. It took a moment to reinterpret the images that his eyes presented, but it was soon clear that he and his constant companion were in a very cramped cabin. It was so cramped in fact that there was quite clearly not enough room in it for either of them. The little space available was entirely occupied by the pilot, and he was not large. Scrooge estimated a standing height of about a meter. He could not be sure of this, as the creature's anatomy was so strange that he was none too sure how tall it might be when upright.

"How is it that we can fit into this cabin when it is so obviously tailored to fit tightly around its planned occupant? How indeed did we get here from our previous vessel?"

Scrooge began to speak with more heat as he considered all his recent experiences. "You have been telling me how you can move only so quickly from one place to another and emphasizing what is and what is not physically possible. Yet I have seemed to move from place to place, from time to time, and even to change my size in ways which must surely be quite impossible. How do you reconcile that with all your strictures, may I ask?"

The Spirit seemed in no way put out by Scrooge's aggressive tone. "Oh yes indeed, you are quite correct. Many of your experiences have indeed been impossible, totally impossible. The clue to this conundrum lies in your words 'seemed to move.' All that you have seen so far and have yet to see are visions: sights and sounds that are illusions, although they convey a message of truth. I am a Spirit, and so my actions and observations are not constrained by base physical law. Whenever I show you a vision the message it carries is true, but it may be viewed in such a way as would not be possible in reality. This is in the nature of spirits such as I. Such behavior, such observation of the physical world from viewpoints which are not themselves physically possible, is common also among certain beings of your own world. You call them 'theoretical physicists.'"

"Well then," responded Scrooge, partly mollified by this answer. "If, as you say, I have but been seeing visions and dreaming dreams, why then has it been necessary to spend so much time and effort in traveling from place to place? Why, for example, did we bother with hour after hour of travel in a flying saucer when you could have produced your vision with no advance preparation?"

The Spirit looked slightly embarrassed, at least as far as Scrooge could make out any expression at all beneath an ever increasing profusion of white hair. He muttered something into his moustache which Scrooge was not quite able to catch, though he did hear the words "corroborative detail, intended to give artistic verisimilitude to an otherwise bald and unconvincing narrative."

While they had been arguing the pilot of the giant spaceship had calmly completed the pre-flight countdown, and the first stage ignited abruptly. Motors with mouths like lunar craters bellowed forth their fiery blasts, and the tremendous bulk moved slowly out from the surrounding wreckage of a solar system.

Faster and faster they moved, thrusting ahead of a fiery tail that could consume worlds. After a time the roar and the pressure ceased abruptly as the first stage exhausted its fuel and was abandoned; an empty shell the size of Australia falling back toward the plundered sun. Stage after stage ignited, burned for a time, and then was dropped behind. Abandoning more and more of itself as it broke free of this succession of initial boosters, the rocket thundered on, traveling ever faster away from its star of origin. Its speed relative to its departure point was still well below that of light but was becoming close enough to show the first hints of the exotic transformations of special relativity. Already, ever so slightly, the stars were beginning to bunch together in the sky as space became compressed along the ship's direction of motion. Ever so slightly, but still imperceptibly, the diffuse hydrogen gas that fills the gaps between the stars was beginning to condense, as the space containing it constricted because of the relativistic contraction of distance.

The last booster stage of the rocket ignited, burned, and fell away. Now only the ship itself was left, a tiny speck compared to its original bulk but the ultimate inheritor of the huge energy that had been expended. As it coasted so rapidly through interstellar space, the ship began to change. Vents which looked like high-tech coal chutes opened around its flanks. Complex coils and antennae protruded from its sides, and small lightning discharges crackled around them. A faint glow appeared in space in front of the vessel as electric and magnetic fields ionized the interstellar gas and guided it to the intakes of the ship's ramjet engines. There the hydrogen fueled an array of fusion motors, and nuclear processes within the engines converted some of this fuel's mass to energy, flinging it out as a flaming exhaust, which drove the machine ever faster forward.

The ship had now burnt virtually all of the great supply of fuel with which it had set out; but it could gather fresh fuel from its surroundings, and this new supply promised to be unlimited. As its speed increased toward light speed, so the density of the fuel supply increased also as the space containing it became ever more compressed. Hydrogen was delivered more and more rapidly to the ship's gaping mouths. That, from the viewpoint of the ship's designers, was the good news. The bad news lay also in the rapidity of the incoming hydrogen fuel. The Spirit explained that the craft was now moving so quickly that the arriving hydrogen atoms struck it with a speed close to that of light. The nuclei of the incoming hydrogen atoms now arrived as dangerous ionizing radiation, a rain of potential destruction as violent as any *Star Wars* weapon, and the ship carried massive shielding to protect the pilot from its effects. On earth radiation comes as charged particles moving very quickly. Here the hydrogen nuclei had been drifting in space, and it was the ship that had come upon *them* with enormous speed. In either case the *relative* velocity was very large, and so the effects were the same.

As the ship continued to accelerate, the *speed* of the incoming hydrogen could not increase much further, but the *momentum* of the colliding atoms could. There was no limit on that. As each energetic atom struck the ship from in front, it gave up its relative momentum and slowed the ship slightly. The atom, once collected, might then be fused in the engines and used to drive the ship forward, but the net gain became steadily less as the incoming atoms struck the more violently.

"What then is the balance between overall gain and loss?" asked Scrooge when these effects were explained to him. "If more and more fuel is becoming available, but when it does it is of less and less use in driving the ship forward, is there or is there not a net gain?" Net gain or loss was something very dear to Scrooge's heart.

"Your question is academic," responded his instructor, "as the contraction of space has become so extreme that the ship has already reached the edge of the galaxy. In the even greater emptiness of intergalactic space the density of matter is much diminished and the craft will effectively be coasting, though still carrying with it a tremendous velocity relative to the galaxy it is leaving."

A thought occurred to Scrooge, rather late it must be admitted. "If this creature is moving so far away at such great speed, how will he or she or whatever return to his, her, or its home?"

"There is no way. Even if a method were devised for the ship to slow, to turn, and to acquire a similar speed in the opposite direction, it would be to little avail. The time dilation effect would be such that the race would long have perished. In this case the Twin has traveled too far and too fast."

Scrooge was aghast. "Was this known to them before the venture began?" he asked.

"Of course it was. I told you that the race had a philosophy much different from yours. But come, we must depart. Though it is impossible for this ship to return to the time and place of its departure, we are not so limited because in a sense we were never here. These examples I show you are nothing but thoughts made visible to you, and we may leave them whenever I will it. All these players, yea the great galaxy itself, as I foretold you, are but visions and like an insubstantial pageant fade, leaving not a wrack behind. They are such stuff as dreams are made of."

As the Spirit uttered these suspiciously familiar sounding words there was a fleeting moment of disorientation, which Scrooge had come to recognize as an indication that the Spirit had become tired of his current vision and was moving on quickly to something quite different. When it was over, they were floating together in the center of a very strange construction. All around them was an array of long rods or cylinders that fitted together at right angles to one another. At each join there was a sphere from which protruded six cylinders to make a uniform three-dimensional structure that looked rather like a wine rack, or a ball and stick model of a crystal such as one may occasionally see in chemistry laboratories. This repeating structure of vertical rods, transverse rods, and longitudinal rods appeared to go on forever, stretching away to infinity as far as Scrooge could tell. The effect was made more fantastic by a gentle yellow glow that diffused from each and every one of the spheres and cylinders.

CHAPTER 6

A Point of View

Scrooge stared in wonder at this repetitively gargantuan likeness of some child's construction kit. "How may such a structure as this exist? It must be huge, enormously larger even than that monstrous rocketship which you showed me. It is inconceivable that any race could construct something on this scale, but neither is it conceivable that such a structure could be natural."

"You speak of monstrous size, but how can you say what is the true size or the scale of this construction? You must remember that occasion when you looked on the world from a small scale viewpoint and saw a fine crack in a rock appear to you as a great canyon.

As was the case then, you have no way to tell whether this scene is as immense as you believe. It might be moderate in size and your time sense so speeded up that you can travel through it, even traveling at speeds very close to that of light, for as long as you wish."

"Well then, is it huge or not?" demanded Scrooge.

"Does it matter?" was his only reply. "Come, let us observe how this structure appears to a moving observer." The vertical and transverse cylinders began to drift past as Scrooge and his companion moved forward. Faster and faster, the drift became a steady flow and then a rapid rush as they sped through a clear space between the surrounding cylinders.

At intervals the Spirit called out their speed relative to the surrounding network, like some airline captain who had neglected to bring his plane. Initially the appearance of the grid showed little change, apart from speeding past ever faster. As his reported speed approached some nine-tenths of light-speed, Scrooge began to notice changes. The vertical and transverse cylinders commenced to bend and curve slightly from the straight and level form they had initially possessed, while the longitudinal cylinders stretched out so that successive junctions seemed farther apart. Their speed increased further relative to the interminable structure, and they rushed past line after line of horizontal and vertical cylinders, stretching away on either side.

Scrooge saw how the Spirit's long hair streamed out behind him as they sped along and was moved to comment. "Surely you told me that it is not possible to tell whether you are moving, and only movement relative to other bodies is meaningful. I feel no atmosphere and no wind, so why does your hair blow out so as we move?"

"Oh that!" replied the Phantom. "That is just to make it easier for the illustrator to show our speed."

Their speed past the cylinders, as announced by the Spirit, rose to ninety-five percent and then to ninety-nine percent of the speed of light. The visible distortions became more and more pronounced. The transverse cylinders bent grotesquely into high, soaring arches, and the longitudinal cylinders that joined successive planes of spheres and rods stretched out farther and farther, so that the more distant parts of the structure receded and shrank into an intricate curved jewel shape before them. This was made the more dramatic by the fact that the color of the rods in front changed steadily from yellow to a deep, vivid blue and glowed beckoningly as they approached.

"But surely this is all wrong!" cried Scrooge. Abruptly their movement stopped, with great disregard for the conservation of momentum, and the Phantom turned to look quizzically at him.

"You think so, do you?" he asked. "To what particular feature do you object?"

"You have told me that the distortion of space and time in relativity means that distances become shorter along the direction of motion, but here we see straight rods appear curved; and the rods that lay along our direction of motion have visibly stretched out far ahead, rather than shrunk. Surely what you have just shown me seems the exact opposite of what you said!"

"You are confusing two different things," replied the Spirit firmly. "I said that you would find that objects moving relative to yourself had become shorter. I did not say they would *look* shorter— that is quite a different thing! You must remember that you see things by means of the light that comes from them, and that light, not surprisingly, travels no faster than the speed of light. It takes time for the light to travel and so you are always seeing things as they were in the past. The farther away they are the farther into their past you see. If you are moving toward something almost as quickly as the light that it emits is traveling toward you, then you can move a long way in the time it takes the light to reach you. This means that more distant objects seem farther away because you are seeing them earlier, when they *were* farther away. Let me give you an analogy.

"Imagine, if you will, that your Victorian ancestor was expecting two friends to visit him. This would of course be in the latter part of his life when he actually *had* friends. Let us call these friends Abraham and Benjamin. They knew that, despite his change of heart, old Scrooge had lived most of his life as a recluse, and so they did not want to come upon him unexpectedly. On their journey each one brought along a box of carrier pigeons and stopped at every milepost to release a pigeon with a note reporting his present position, so that Scrooge might be adequately forewarned of his arrival.

"At one moment two pigeons would arrive. The one from Abe would say that he was now three miles away, the one from Ben that he was six miles off. A little later another two would arrive. Abe's would say that he was now two miles away, Ben's that he was five. From this you might assume that they were both riding steadily toward their friend, separated from one another by a constant distance of three miles. If you thought so you would be mistaken. "Consider more carefully the implication of these messages. Let us view these gentlemen from a high vantage point, which allows us to see their movements."

The strange array of rods and spheres that had surrounded them faded from sight, and Scrooge found himself floating in the air high above a normal earthly landscape, much as he had earlier in the company of the Shadow of Entropy. Below he could see a long, straight road down which two horsemen rode steadily, stopping briefly at every milepost to leap off their horses, pen a short note, and send it off attached to the leg of a pigeon; then once again continue on their way. By comparing the positions of the mileposts Scrooge could see that they were both riding at the same speed but were *two* miles apart, not three.

"There you see the two friends," commented the Spirit. "Abe is riding two miles ahead of Ben, and both ride at a steady ten miles per hour. At this speed they pass a milepost every six minutes. Their pigeons can fly at thirty miles per hour; so when Ben releases a pigeon it flies on ahead, eventually overtaking Abe. If Ben releases a pigeon when he reaches the six-mile post, Abe will at this time be

beside the four-mile post. The pigeon will catch him six minutes later, but by this time he will have reached the three-mile post and be releasing another pigeon. It is in company with this pigeon of Abe's that Ben's pigeon will fly the remaining distance to Scrooge's house and so he will always receive together two pigeons which report his guests as being three miles apart when in fact only two miles separate them. The pigeons *have* come from points three miles apart, but *not at the same time.*

"It is the same with light. The light from more distant objects has started out earlier, when they were farther away, and so distances appear to be stretched out. This is the main effect that distorts what you see. If you allow for the length of time it takes the light to reach you, you have a measure of where things *actually are*, from your point of view. You will *then* find that lengths have contracted as described by relativity, but that is not the most obvious effect that you will actually *see*. The fact that light takes time to travel explains why the longitudinal rods have stretched far into the distance and also why the transverse rods have curved. The center of a straight bar that crosses your field of vision will be closer than are its far ends, so these ends will appear to curve farther away from you."

"That is all as may be," Scrooge responded rather grudgingly, "but how do you explain the change in color that I saw? Surely that has nothing to do with the speed of light."

"Well actually it has. That is an example of what you call the Doppler effect. Light travels as a form of a wave, rather like ripples traveling on the surface of a pond into which you have dropped a stone. Such ripples have a wavelength: the distance between successive peaks in the disturbance. They have a frequency: the number of times per second that the water at any point on its surface will undulate up and down. Waves have a velocity and this is the speed with which the ripples may be seen to travel over the surface of the pond.

"We know that the *velocity* of light is constant, but the wavelength and frequency need not be. If you move toward light which is coming toward you, then you are plowing into the incoming waves and you meet successive peaks more quickly. You see the light with a shorter wavelength and a higher frequency. Blue light has the highest frequency of any that you can see, and so light becomes more blue when you move toward its source. If you move away from the source of the light, then the opposite result is produced. The wave peaks are farther apart, the frequency with which you meet them is reduced, and the light is red-shifted. This is much like to the effect you hear when an ambulance passes you with its siren sounding. As the siren approaches you hear a high frequency, which falls abruptly as the ambulance passes; and as it races away from you, you hear a lower frequency.

"It is necessary to distinguish between the view that you will *see* and the actual position of objects in your own frame of reference. To gain a more realistic picture of the world you need to look at it in four-space. Allow me to show you."

Scrooge looked around him and found he was standing upon a wide "plane." These were no clues as to its whereabouts because the sky was featureless, and no sun or stars were visible. The sky was not simply overcast with heavy cloud, it was totally blank as if the creation of this landscape had not bothered to include the sky as a feature. All around him, boundless and bare, the lone and level plane stretched far away. Beyond a certain distance it became impossible for Scrooge's eyes to resolve any detail upon it, but as far as he could tell it remained quite flat as it stretched away to infinity.

There were no familiar features, and Scrooge found it quite impossible to judge the scale of the scene before him. He saw that the plane was marked with a pattern of concentric circles, each one being surrounded by another larger, yet to form a target pattern whose outer limits stretched away farther than he could discern. With a sudden feeling of acute paranoia, Scrooge noted that the target on

☾ DOPPLER RED-SHIFT FOR STARS ☽

Though the speed of light does not change, no matter how the source or observer may be moving, the frequency and wavelength *can* change. If the source is moving toward you (or you are moving toward the source, it is the same thing), then the waves come closer together and the frequency is higher. The color of light depends on its frequency, and this increasing frequency moves the color toward the blue end of the spectrum. If the light source is moving away from you, then the effect is opposite and the color is shifted toward the red end of the spectrum.

The line spectra (see Chapter 11) from distant stars show pronounced red shifts, suggesting they are all moving away. The farther the distance to the star, the greater is this red-shift. The universe as a whole appears to be expanding, with every part moving away from every other.

which this whole monstrous pattern of concentric rings was centered was the very point where he himself was standing. Without pausing to analyze further the possible implications of the scene, he ran rapidly to one side so that he might remove himself from this uncomfortably prominent position.

After sprinting for some distance he paused and again looked around. The strangely sterile landscape was as before. Indeed it was *exactly* as before. The entire pattern of circles was still centered exactly on the point where Scrooge now stood, panting to recover his breath. This observation did nothing to diminish his feeling of exposure, but before he could become seriously worried he was distracted by the inevitable voice of the Spirit.

"Here you see a representation of the way that you observe space. Each circle labels points in space that are equally far from your position. You have obviously guessed this already, since I saw that you took steps to confirm that the pattern remained centered on wherever you might be.

"As you look around you in space you see events. You do not just see things, you see them as doing whatever they may be doing at the time you see them. They may not be doing anything very interesting, they may not even be doing anything at all that you can detect. Whatever is the case, that is what you see. At any given time, you see the actions of that moment."

As the Apparition spoke these words, the landscape became filled with tiny scenes of activity. Scattered here and there over the plane was a host of tiny manikins. Little featureless doll-figures, they were each engaged in some form of activity. They cooked, they slept, they chopped wood or sowed seeds. Some fought, some bound another's wounds. Many of them just sat around doing very little. Each had its own activity, but as Scrooge watched these would change. The wood-chopper would light a fire, the cooks would serve and eat the food, the warriors would prevail, or perhaps they would die. Some of those sitting around would began a more purposeful activity. Everywhere there was change.

"At each moment you see different happenings, different events. As I have already pointed out, the events you see at any given time are not those happening at that time in your frame of reference. If you see something happen a distance away, it can only be because light has traveled from the event to your eye. It takes time for the light to travel, and so what you see is what was happening when the light set out, not what is happening *now*. Your eyes are telling you old news.

"In your normal affairs the things you see are close at hand, and the time it takes the light that comes from them to reach you is so short that it hardly matters. For sufficiently distant objects, however, the time it takes for light to travel to you may be long. The light from your Sun takes about eight minutes to reach you; from the nearer stars it can take many years, while light from more distant stars may take thousands or millions of years. You could say that you see those as they were a long time in the past."

The Spirit stood a little way off to his right. Scrooge noted this and noted also, with a slight recurrence of his first paranoid feelings, that the other was standing well away from the center of the marked circles and that they had not followed *him*. It was clearly Scrooge, and Scrooge alone, who was being thus singled out.

"This plane, which you see stretching around you on all sides, is a representation of space. For convenience, in the creation of this vision three dimensions have been shown as two. You must use an analogy and imagine that all of space is contained in this flat plane that you see around you. Limiting the number of dimensions makes it that much easier for you to understand the way that space and time intermingle to make four dimensions. It is easier to visualize two dimensions being extended to three than it is to imagine how three dimensions would look when extended to become four. It is difficult in the extreme to visualize the eerie stretching of space and time as they blend together. For that you would need the gift of four-sight."

Scrooge was upset by this comment as he had always felt that he possessed at least his fair share of foresight, and that it was not on many occasions that he was caught unprepared. He had to admit, though, that his present experience was certainly one of those occasions.

The Apparition paid no heed to his reaction. "In order to see how space and time work together we must add to the dimensions of space the extra dimension of time. To describe space you talk of three dimensions, which are given by three coordinates at right angles to one another, right or left, backward or forward, up or down. We must add to this a dimension of time, which is also in some way at right angles to each of the previous three and extends between past and future. Then time is a fourth dimension, but in our reduced representation it is only a third direction and that you can imagine. The dimension of time is not really the same as the three dimensions of space. There are important differences, but there are important similarities as well.

"Now we shall extend our vision to include time, measured up or down from our plane of space. In order that time may be shown on the same scale as distance and plotted along this direction, it must be multiplied by the speed of light. It is on this scale of speed that space and time show their remarkable properties; and because the speed of light is so large, a little time will correspond to a great extent of space. The level plane around you has changed its significance. It currently represents your true *now*: the events that all occur at the same time, at least as *you* judge time. This means that the different events that you *see* at the same time will no longer lie on the plane, as they must actually have happened earlier."

The pattern of concentric rings sank below the ground, and at the same time the ground became transparent. Perhaps it always had been so, as before there had been nothing to see below it. The smaller, closer circles sank but a little distance, while those farther away sank proportionately further. Now all of the circles lay upon the surface of a long, spreading cone that extended below him, beginning at the point where he stood and stretching far down into the infinite past. When he raised his eyes Scrooge could see *another* cone, twin to the first, which started also from the same point at his present position and opened out into the future above him.

"That is the light cone," remarked his companion. "On its past section lie all the points in space-time from which light may reach your present position, everything in space that you may see at the present moment. The future half of the cone contains the events that might receive light that is leaving your position. On the past cone is everything which you can possibly see now. On the future cone is everything that can possibly see you as you are now. The light cone is a personal thing, you see. Everyone has his or her own. More accurately, every *event* in space-time has its own light cone. You share the use of the one that is centered on your position in space and time, and every other event is separated from you by some *interval*.

"What do you mean by 'an interval'?" asked Scrooge. "Should it not be a distance or a time, depending on whether the point is distant in space or in time?"

"An interval has the nature of both," replied the Spirit. "In general, events will be separated from you in both space and time, and the two combine together to make an *interval* in space-time. As time is considered to be rather like another dimension of space, it is included to find the length of an interval in much the same way as the three separate dimensions of normal space combine together. The relative position of any two points in space is given by quoting the differences in the coordinates measured along some three *axes*, three directions at right angles to one another. These three numbers give you the separation of the two points. If the points lie on the same left-right axis and one of them is some distance to the left of the other, then obviously that is the distance that separates the points. If the second point is a little distance to the front of the first, as well as being to its left, the separation is a little greater. Wherever two points are separated by distances along two or more of the three coordinate directions which describe their positions in three-dimensional space then their overall separation may be found from *Pythagoras' theorem*. The square of this separation is given by adding together the squares of the individual separations for each coordinate. The effect of having differences in more coordinates is always to *increase* the separation of the points. Adding together more squares can never make the result less since the squares of distances are positive.

"The time coordinate also contributes to the sum an amount equal to the square of its magnitude, but the way that time is included is different. When there is a difference in the times for two events, the square of the time difference is subtracted from the space part of the sum, not added. This follows from the rather peculiar relation between space and time which arises from the transformations of special relativity. Because the time part is subtracted it follows that the square of the interval between two events in space-time may be positive or negative or even zero, depending on the relative sizes of their separations in space and time."

"If this interval is something like the distance apart of two points in space," began Scrooge musingly, "then events that are separated by a zero interval must surely be the same event, must they not? If two points in space are no distance apart then they are of necessity the same point."

"It is rather different for space-time," explained his instructor. "Two events can be separated by an interval of zero length and still not be the same event. The interval between any two points on the same light cone turns out to be zero, and for this reason the light cone is often known as the *null* cone. Actually, in a strange sense, you *could* say that these are one and all the same point. Near the light cone the dilation of time is such that anything that travels along a path lying very close to the light cone would find that little time had passed, and an equally extreme compression of space would mean that all distances would shrink. Along any path that lay along the light cone itself these effects would reach their ultimate conclusion and, to an outside observer, time would effectively be frozen.

Everything along the path would occur at the same time. An observer traveling along such a path—a path which would require that he be moving at the speed of light—might not find that time was frozen since relativity prevents an observer from seeing any such effect of his own motion. The surrounding universe would, however, be completely contracted so that all of space would flash past in a blink of an eye and *this* would mean that for him all events in the outside universe would occur at the same time."

The Phantom indicated the double-ended light cone with a sweep of his arm.

"You can see that your light cone divides space-time into a number of distinct sections. First there are those events which are *inside* your cone and those that are *outside*. Points within the light cone are separated from you by intervals whose square is negative. This is strange to start with because the square of any normal number is always positive, whether the number itself is positive or negative. A distance whose square is negative is thus a little peculiar. In fact it is more like a time than a distance. The intervals within the light cone are consequently called time-like. Along the center of your light cone lies the time axis itself. This is the path through space-time that is followed by anything that is at rest relative to you. The light cone has two distinct halves."

The Spirit pointed to indicate the two conical boundaries in space-time whose vertices touched one at the point where Scrooge was standing. "Ahead of you is the half that contains your future; the half behind you contains your past. The two halves of the light cone meet at your present, your *here and now*. Away from the axis, but within the light cone, are the events in space-time that could be reached by starting where you are now and traveling with a velocity less than that of light. The greater this velocity, the closer to the light cone will lie these points. The light cone itself marks a limit. On it lie the points that can only be reached by traveling at the speed of light, and nothing can go faster than this."

With a further sweep of his arm the Visitation indicated that whole volume of space-time that lay completely outside the light cone. "Outside are those events that are separated from you by an interval whose square is positive. Normal distances have a positive square, and these intervals correspond to normal separations in space or, at any rate, something that is more like a normal separation in space than it is like a time interval. You must keep in mind that we are not just talking about space here but about intervals in the combined extent of space-time. In general, these will be a mixture of

᭦ SPACE-TIME AND THE LIGHT CONE ᭥

Time seems to behave in some ways like the dimensions of space. This means that the idea of three-dimensional space may be extended to give four-dimensional space-time, now including an axis of time. Locations in space-time are called *events*, as they specify both the position *and time* of a happening.

It is rather difficult to draw diagrams in four dimensions, so the convention is to show all of normal space on a plane with only two dimensions and use the third dimension to represent time. Space-time is divided into three sections, all centered on the observer's *"here and now."*

The division is made by the *light cone*, which is the path that light would take to or from the observer and marks the limit for the exchange of energy or information.

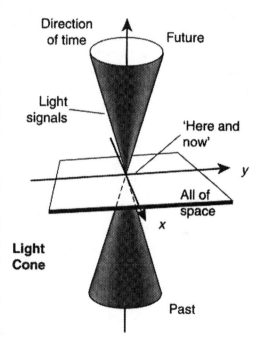

The light cone is divided into two parts.

The *future cone* contains all of space-time, which may be reached by a signal from the observer's "here and now." It contains all events on which it is physically possible for the observer, starting now, to have any influence.

The *past cone* contains all events from which a signal could reach the observer's "here and how" and which might have an influence on him.

Events outside the light cone are separated by *space-like intervals*. They are completely out of the observer's reach.

space and time; but when they have a positive square they are, on the whole, more space than time and so are called *space-like*.

"Any points with a space-like separation are *totally* isolated from one another. You *cannot* get from one to the other. No signal or energy of any type whatever may pass from between them, even if it travels at the speed of light. This is not to say that there are points in space that are so far away that they cannot be reached eventually, but you must remember that events in space-time have both a position *and* a time, and what I am saying is that it may not be possible to reach that *position* before that *time* has passed."

The Phantom made yet another wide, sweeping gesture, this one encompassing the entire surrounding plane. "This flat plane, you will recollect, is to be taken as representing the three dimensions of normal space. This space is no longer just what you can see, it is actually *your* NOW. The events which lie in it are those that you would calculate to occur simultaneously, allowing for the time it has taken the light from them to reach you."

"That sounds reasonable enough," declared Scrooge. "Why, though, did you stress that it represented *my* now. If different things should happen at the same time then at that particular time they will all happen 'now' for everyone. Now is now, and it has nothing to do with me!"

"Oh, but it has! Things that happen at the same time for you will not necessarily happen at the same time for others. It is all relative. Look at the surrounding map of space-time. "I have told you that the central axis of the light cone is the path of your future along which only time is changing. This direction in turn defines the plane of the present, which contains all of space at the time you call now. Consider the future for some other observer, one who is moving rapidly away from you. As time goes on he will get ever farther away, and so the path of his future will incline at an angle away from yours.

"The path of his future is that line along which, for him, only time changes. It is the route followed by his own frame of reference— that frame in which he is of necessity at rest. As you have your set of directions in space-time, so does this other observer. The direction along which he sees that time alone as changing will define the plane of *his* present, for which time does *not* change. Since time and space are intertwined, the plane that contains *his* present will tilt relative to yours because his time axis is inclined to yours. His set of axes will be *rotated* relative to yours, but because this is space-time rather than normal space the rotation is a little unusual. If we were looking at perpendicular axes in ordinary space and the vertical line were to rotate clockwise so that it swung toward the right, then the plane at right angles would rotate clockwise also and would dip down toward the right.

"Space-time is different. I told you that space and time had to make strange contortions to keep the speed of light constant, and this is one of them. Though time is like a dimension of space, it is not a *normal* dimension. In fact, the time axis is imaginary."

At this Scrooge exploded passionately. "I knew it! I knew it! At last you have admitted that this is all just make-believe, with no resemblance to reality! An imaginary axis indeed!"

"Calm yourself," commanded the Spirit coldly. "It is but a name, a name that mathematicians of your day have given to those numbers that are negative when multiplied by themselves. It is a strange property, certainly, but not by any means unreal. It is unfortunate that such numbers are described as imaginary, but it would be better to think of them as *imaginative* numbers, numbers that have the vision to do what ordinary numbers find to be impossible.

"An effect of this 'imaginary' nature is that if the time axis rotates in a clockwise direction it causes the space plane to rotate in an *anti-clockwise* direction. As the speed of the moving observer approaches that of light, the time axis and the space plane will approach one another and eventually become the same. It is really just as well that it is so difficult to travel at speeds close to light speed."

The scene around Scrooge changed to include the space-time axes suitable for a new observer. To make the distinction clear the sets of axes were color-coded. A red line sprang into existence at an angle to the right of the central axis of Scrooge's light-cone. At the same time a red-outlined section of the plane swung up toward it. This was the plane of the new observer's NOW. On it lay all the events which that observer would reckon as happening at the same time. Since the new plane was inclined to Scrooge's original "plane of the present," events that had all fallen upon *that* could not lie also on the new plane. Events which he saw as occurring at the same time were at different times for this other observer. The farther away the event was in space, the farther into the new observer's past or future was the event.

"You see there a consequence of relative time. If time were totally independent of space, as it was once believed to be, then the times at which events happen would not depend on the speed of the observer. In fact they do. We now know that time is not totally separate from space; the two have become entangled. You may say that two events, with which we shall follow tradition by calling A and B, are happening at the same time, and for you they would be. If another observer were moving relative to you with a speed close to that of light then he might say that was *not* the case, that in fact event A happened before event B. Yet another observer might be moving in the opposite direction, and she would say that you were both wrong and that A happened *after* B."

"And who would be correct?" asked Scrooge.

"Why you would all be completely correct! Of course you would! Each would have correctly stated the truth as it appears to him or her, and what more can anyone do? There *is* no absolute time. There is no absolute NOW unless it is the *here and now*. Events that occur at the same time *and* the same place are the only ones that are at the same time for all observers.

"Look far into the depths of space!" cried the Spirit abruptly. Scrooge looked and saw different events which chanced to happen at the same time but at different points across the vast depths of space. It was perhaps not correct to say that he "saw," as he was now aware of events at the time they occurred. The great spans of time needed for the light to travel to his eyes from such distances had already been adjusted out by the workings of the Spirit's vision so that, as far as the words have meaning, he saw them *when they happened.* Call it what you will, when he looked in one direction he saw far away an asteroid collide with the surface of a distant planet. He looked in the opposite direction and saw a star suddenly burst out in an explosion of light as it entered a nova phase. He looked between and saw, less dramatically, a long jet of hot plasma rising from the surface of a nearer sun.

"Now see how those events seem from a moving frame!" Scrooge and the Spirit soared away and then came rushing back across the landscape at phenomenal speed. So at any rate the Spirit assured him. It was difficult for Scrooge to say of his own knowledge that he was moving much differently, since he was still in the center of his own light-cone, and his uniform surroundings gave no reliable reference points by which he might judge motion. He was directed to look again into the far distance and saw the startling outburst from the nova. He saw also the nearer sun, but now it was glowing steadily and serenely with no sign of disturbance. He saw the asteroid, coasting along in an orbit that looked dangerously set for a collision with a nearby planet.

Some time later, he could not really say how long, he saw a glowing shell of gas, which was all that remained of the exploded star. He saw the jet of plasma erupt without warning from the calm surface of the sun. He saw the asteroid unmistakably bearing down upon the unfortunate planet. Later again the remnants of the nova were visibly expanding and cooling, the glowing prominence sank back into the surface of the sun, and the asteroid became a flaming meteor tearing through the atmosphere of the stricken planet.

"And now the events when you are moving in the opposite direction!" Once again Scrooge and his companion soared away and then came rushing back, but this time from the opposite direction. Now when Scrooge looked around he saw a dying star, meting out the last energy from its exhausted nuclear fires. He saw the sun, again glowing steadily and serenely as it had before the plasma erupted. He saw the asteroid crash into the planet and tear through its atmosphere, reminding him of the occasion when the Mistress had sent him to accompany a much smaller meteor in its fiery descent. Later he saw a huge expanding ring of smoke and water vapor: a planet-wide mushroom cloud that hid the devastation below. He saw the jet of plasma erupt without warning from the calm surface of the sun. The dying star had exhausted its nuclear fuel and was on the point of final collapse. Later again, his final view showed the star abruptly collapse and then flare out monstrously in a nova explosion. The glowing prominence sank back into the surface of the sun, and the planet stricken by the asteroid was now all but totally obscured in a thick mantle of cloud.

"And there you see it," concluded the Phantom. "You have seen how the same events may be seen to occur in a different order as they are seen by people moving at greatly different speeds."

"Perhaps I have seen it or, rather, the vision that you have chosen to show me, but I find it hard to believe that my viewpoint, the way I am moving, can so affect what I see!"

"It is because time has become in some ways like a direction in space that you find that such things depend on your viewpoint," replied the Spirit. "In your normal three dimensions it does not surprise you that the appearance of something should depend on your viewpoint, on how you look at it. If you look at a ruler from an oblique direction in space you will see it as foreshortened and if you look from a fast moving reference frame this will give you an oblique viewpoint in space, time and so you see foreshortening. Such foreshortening manifests itself as the relativistic contraction of space. The other effect that you have just seen, the way in which the sequence of two events in time depends upon the velocity of the observer, this also is akin to an effect you can see in your normal three dimensions: the parallax effect that you observe as you move around."

Abruptly and without warning the scene changed, in a fashion that Scrooge had become to recognize as indicating that the Apparition wished to make some point but would not take the trouble to build up gradually to the transition. Scrooge was now standing with his companion on a low, grassy hill in the middle of a country scene, reminiscent of a typical work of the major school of English landscape painters. The far side of the hill was planted with a scattering of small trees. Scrooge noticed that one tree with a particularly straight trunk happened to be directly in line with a distant church steeple.

"You will see that the church steeple is directly in line with one of the trees," remarked his companion. "Pure coincidence of course. Observe what happens if you take a few steps to your right." Scrooge did as suggested and saw that the tree and steeple were no longer aligned, the distant steeple being now to the right of the tree. On the Spirit's further suggestion, he then walked across to his left and saw, as one does, that the steeple had apparently moved and that it was now to the left of the tree.

"There is no mystery in any of this," said his guide. "The phenomenon of parallax must be quite familiar to you. As you move, so your viewpoint gives you a different line of view. From one position the line may pass through both tree and steeple. From a position to the right your line of vision will be from a different direction and will pass to the right of the tree. If you had moved to your left it would pass to the tree's left. In space-time your 'line of vision' will include time as well as space. A different viewpoint is provided by fast relative motion, and when you look at distant objects from such different viewpoints you may see the more remote objects on the 'future' side of the nearer ones or on the 'past' side, depending on your viewpoint, that is, on your direction of motion."

"But this cannot possibly be true!" protested Scrooge. "There can be no ambiguity about the order of events. I cannot eat a meal before it has been cooked. I cannot finish a letter before I have started to write it. Effect follows cause, and this can never be a matter of opinion or viewpoint!"

"I have not said that it may be. You may have noticed that the events you saw were obviously quite unrelated to one another. They might just as well happen in one sequence as in another. There was no question of cause and effect being in any way involved between them." Scrooge admitted that he had gained that impression.

"Events that may be seen in different sequence by different observers *must* be unrelated to one another! That is the whole point! The sequence of events can depend on the motion of the observer, in the manner I have just illustrated, only when the *separation* of the events is *space-like*. That may clearly be demonstrated. There can be no connection, no communication between events with a space-like separation, but it is only such events that you may possibly see as happening *at the same time*. I have just explained that. If two events have a space-like separation, then they cannot be cause and effect because nothing can travel between them in order that one might affect the other. Even light cannot travel from one to the other, and nothing moves more quickly than that. In no way, no way whatever, can one such event affect the other."

"But events do have an effect on one another!" protested Scrooge. "Surely you do not deny the existence of cause and effect!"

"Oh certainly," responded the Spirit readily, "but if one event is the cause of another, then the separation of the two must be time-like, for obviously there must be some way for the event that is the cause to transmit information to the event that is the consequence. When the separation of events is time-like, when it is possible for someone or something to travel from one event to the other, even though they might need to travel very fast indeed, the sequence of those events *is* unique. There is no ambiguity. One is in the future of the other and is so *for all observers*. That means that no observer can see them as happening *at the same time*. If any observer sees two events at the same time, the events *must* have a space-like separation.

"Where the order of past and future really matters, where there is at least the possibility that one event might be the cause of the other, then all observers would see the same sequence in time. You can legitimately ask 'what will happen to me next,' and the same answer would be given by any observer in any frame of reference. What you may not ask is 'what is happening NOW on a distant star?' and expect a definite answer. The answer to such a question must be 'it depends who is asking.' In the words of the old saying: 'There *is* no time like the present.'

"Your future is the sequence of those events that occur at different times in *your* position in space. They make the path of your future. A path is a continuous sequence of places. Grain after grain of sand packs the surface of the road and makes up the totality of the way, each grain being positioned a little closer to your destination. Your life-line is a sequence of events that form the grains of your life. It is your future alone or, rather, the future of the point in space time which you occupy. Time, like space, is relative. But come, let us return to a normal view of space."

As Scrooge heard this he looked around and realized that at some time during the Vision's remarks they had returned to his abstract 'four-space.' Scrooge stood again at the center point of his light cone. It was of course just one of an infinite progression of such cones, each one centered upon a different NOW. The line of his history stretched by him, from his past to his future. Along this line the events of his life passed him like little beads or grains of sand; strung along his life-line they drained from the cone of his future, through the narrow confines of his present and fell away to be lost in the lower cone below him, receding steadily into the distant past.

He looked at this double cone which contained his past and his future and at the grains of time tumbling past. As he watched he was amazed to see a gigantic hand appear that closed upon the double cone and grasped it at its center. The cone was lifted away from him and the hand shrank in his vision while the past and future cones became the upper and lower bulbs of an hourglass, firmly grasped in the bony hand of a very old man with a very long white beard. Scrooge recognized this as yet another aspect of his com-

panion Spirit, now in the traditional semblance of Father Time. Together they stood upon an empty landscape, with no buildings or roads in sight, only rocks, grass, and a few small trees. Dreary though the scene was, it was obviously a landscape that he was looking at in normal three-dimensional space.

Scrooge turned to the ancient figure of Father Time and made a last attempt to voice what he saw as a protest of common sense against an excess of theory. "Surely you cannot be serious in what you are telling me. Surely there must be a time that is uniquely NOW!" protested Scrooge. "You, of all Spirits, must be aware of a present moment that carries us forward as Time flows irrevocably on."

"Wherever do you get such ideas?" wondered the Spirit querulously. "Have I not shown you that time is a coordinate, much like the coordinates of space? Time does not flow, it just *is*. Would you say that space flows? If you were to look at a road, a Roman road that runs straight and true from left to right, would you say that it flows?"

"You have just said that it runs!" Scrooge pointed out, rather smugly.

"Well yes, that is the common expression," admitted the Spirit irritably, "but no one believes that the road is *actually* running. You might just as well say that the road runs from right to left as from left to right and everyone knows that in fact the road does not run. It just lies there with different parts of it filling the distance between two towns."

To illustrate the point, Scrooge found that he and his companion were now standing upon a narrow road, which ran (if we may be allowed the word) straight into the distance upon either side. The Ghost pointed to a spot near their feet. "You may choose any point you wish along the road and call that point HERE. Once that point has been chosen then other points on the road will be on the left or on the right. If you had chosen HERE to be a point some little way farther to the left, then points that had otherwise been close by on the left would now be to the right of this new point. Right and left are relative. In this sense past and future are relative positions. As I have just told you, possible movements of the observer do not affect whether a point in space-time is in the past or future *of any other point* on which it might conceivably have an effect. Whether a point is in *the* past or *the* future is a different thing, because this depends on the point in time that you choose to call NOW. This is a personal thing.

"I have said that there is no universal NOW. For you, as a conscious observer, there is a time which you choose to call NOW. When you talk to your friends and acquaintances they are, of necessity, fairly close at hand. You will consequently be speaking to them at a time which is as close to *your* NOW as makes no difference. Other people must be at points in space which are not the same as the one you occupy, so strictly speaking you have no common time. Nevertheless, they will be so near that any indefiniteness is *far* too small to be noticed.

"There is no *flow* of time involved in this. History, personal or national, is a matter of happenings, of events. Each event in history has its place and its date. Your personal history is a string of events—so many events that to distinguish them you must quote time considerably more accurately that is usually implied by the word 'date.' At a certain time and place you were born, and that is one of the points which is joined by the pathway of your life. At the other end of the path is your death, but in between is a series of events, one after another, along the pathway. Some are major events and marked by milestones. When you pass an important examination, when you get married, when you have a child—all these are special occasions. Between such momentous events there are others. You wish to eat, so you sit at a table in front of a plate of food. At some point in time and space your hand reaches for a fork. At another point, close by in time, your hand closes upon the fork. Near this in space-time are the events where you lift the fork, where you use it to pierce a sausage, where you lift the sausage toward your mouth. At the end of this sequence you bite the sausage and begin to chew, but other events follow. The road between the milestones stretches on and on and is close-packed with a continuum of events, one after another without clear distinction."

The surface at their feet had assumed a slick, plastic appearance, with dotted perforations which were down the edges, rather than the middle, of the 'road.' Down the middle ran a succession of frames

containing pictures that showed the sequence that the Spirit described. Each picture was still and complete, a frozen moment but giving the illusion of time passing when all were viewed in succession. Down the road of time ran this still succession of 'moving pictures.'

"Events blend one into another, and each has its location in space and in time. As you travel a road you become familiar with the portion you have traveled, but the road ahead is unknown to you. So with your life. The part that lies behind, the part which you call the past, is known to you, assuming you have a good memory. The part that still lies ahead is largely unknown, but you may be sure that your future also is a sequence of events spread out along the direction of time, unknown though they may be.

"The future is not entirely unknown, however," continued Father Time briskly. "Because the future is a consequence of the past you can often make a good guess at what it may contain. When you put the sausage in your mouth you may confidently expect that you will eat it. This thought brings us to consider not just the sequence of time itself but what actually happens *in* time. The essential feature of time is that it allows the possibility of change. The distinction between past and future is only significant because things do change and you expect the future to be different from the past. You could say that time has no meaning unless things do change, since change is the observable effect of time, and so it is to change, to *cause* and *effect*, that we must now turn our attention."

CHAPTER 7

A Clockwork Universe

S tanding four-square in the middle of the barren landscape, the ancient be-whiskered figure of Father Time continued to hold forth on the nature of time and its consequences. "Time allows change and, at its most evident, change means movement, which is the subject of the science of mechanics. You have already met with some of its important concepts when you were visited by the Mistress of the World and she spoke to you of energy and momentum. These conserved quantities put constraints and limitations on how things may move,  but as a general rule you may say that the way an object does move is changed by applying a force. If the object were stationary the force would make it move. If the object was already moving, then the force may change the way in which it moves or perhaps even bring it to rest. In the absence of any forces, an object will continue to coast along in a straight line with whatever velocity it happens to have. As uniform velocity is purely relative to the observer's frame of reference, you have no cause to say for certain that the object has an absolute velocity. Any object has the right to consider itself to be at rest if it feels no acceleration. It is changes in motion that make for interesting behavior, and these come from the application of forces. But what are forces?"

Scrooge had been asked this question before, and this time he wisely decided against answering what was obviously a rhetorical question.

"Whenever a force acts there is always something which actually produces the force, which is in fact the *source* of the force. One of Newton's laws says that 'for every action there is an equal and opposite reaction,' which is to say that a force will always act both ways. It will pull or push its source just as much as the object that it acts upon. If you fall toward the earth you may say it is because the earth exerts a gravitational force upon you that pulls you toward it. It is equally true to say that you exert a gravitational force upon the earth that pulls the earth toward you. The forces are as equal as is the truth of these two statements. The action of the force that makes you fall is balanced by an equal and opposite reaction that you exert upon the earth. You would both be accelerated by the force, but because the earth is so much more massive than you are its movement would not be obvious. The earth and you would acquire equal and opposite momenta, however, and the total momentum would be unchanged. Newton's law and the conservation of momentum turn out to be the same thing.

"In such cases you may see that the force is affecting two things quite some distance apart. You might not be very far from the *surface* of the earth, but you are a long way from its center, and every part of the earth exerts some force. The separation of the two bodies that are affecting one another is more evident when you consider how the Sun's gravity holds the earth in an orbit about it."

"Yes, how does it?" asked Scrooge. "I do not see by what means the Sun is able to affect something so far separated from it. It seems to require some form of action to take place between objects that are a great distance apart, one that allows the Sun directly to affect the earth without the assistance of an intermediary."

"That is a question which naturally arises. The idea of 'action at a distance' is not favored by scientists, who prefer to think that there is something which does connect the bodies concerned. They believe that there is a *field* occupying the space between the Sun and the earth, and that this field transmits the force over the distance between them. Relativity says that no effect may be transmitted between two points more quickly than the speed of light, and that this limitation must apply also to the force of gravity. No one can deny that gravity produces an effect.

"If we say that the speed at which the effect of the force can travel between its source and destination is limited, then it seems reasonable to believe that *something* is in fact traveling at this speed through the space between, and the effect is not just *appearing* a long way away, while skipping the intermediate gap.

"The theory of special relativity tells you that no energy can travel more quickly than light. This puts a restriction on the speed of spaceships and radio waves as a means of communication between planets that are in orbit around different stars. It is a statement about space and time, not about light as such; and the same limit applies to any means of communication. One frustrated writer of science fiction came up with what seemed a splendidly straightforward device for sending messages faster than the speed of light."

The ancient figure stretched out his hand, and Scrooge saw that he was holding one end of a long rod. A very long rod it was, extending up into the sky and vanishing from sight. It was so long that its far end could not be seen at all.

"What is that?" asked Scrooge. "It appears to be just a long rod."

"That is what it is, a long rod. The device was a long, stiff rod, nothing more than that. It is, however, a very, very long rod, several light years in length. The idea was that this rod would extend from one planet to another. Information might be sent by jerking one end of the rod to and fro, whereupon the other end would tap out a message in Morse code."

The Spirit paused briefly. "There are several practical engineering difficulties with this scheme," he said dryly, "but the basic idea would not work anyhow. If such a rod could be made and one end were moved, this movement would not be transmitted to the other end more quickly than the speed of light. The atoms of the rod are held in their relative positions by electrical forces, and there would be no advantage at all in having the material of the rod fill the gap between the stars, instead of relying solely on the electrical *field* that is involved in the transmission of light or radio waves. The disturbance would move through the rod at a speed well below that of light. That is assuming you could move the rod at all, since at that length it would assuredly be *heavy*!"

The long rod in the Phantom's hand broke apart into a twinkle of tiny dust motes that resolved into a host of little spherical particles moving about, bouncing off one another or veering aside to swing around, deflected by some force not visible to the observer.

"Whatever the nature of a force may be, where you have forces you see the operation of *Newton's second law*. It says that the rate at which the momentum of any object changes is simply equal to the force it feels. This is a basic law of mechanics, tested again and again, so many times that it seems likely to be true without exception. If you know how an object is moving, if you know its position and its velocity, and if you know the forces that act upon it, then you can use this law to estimate how its motion will change. Knowing this change, you know what its position and velocity will be in the near future. "If you know the positions and velocities of a self-contained group of many particles, then this knowledge defines also the forces that act between them because forces act between pairs of particles and depend on their separation and velocities.

"When you are pulled toward the earth it is because all the component parts of your body are being pulled by every tiny grain of matter in the earth. Those grains on the ground close to you are pulling more strongly than those scattered over the plains of Australia at the other side of the earth because these latter are more distant, but all will give their contribution to the total. "The positions, the movements, and the forces acting upon the various particles at any time will determine how their velocities will change, and their new velocities will decide where they will be a moment later. Their new positions will give new values for the forces which they experience, and again you might predict how they will subsequently move and where they all will be a little later again."

Bright, glowing lines sprang into view between the different particles, which were still dashing about in front of Scrooge. These formed a cat's cradle of lines between them, a dense mesh that revealed the forces between each pair. Some of the lines were much brighter than others, showing that force to be much the stronger. In general, the shorter the distance between two particles the stronger was the force they exerted upon one another.

Among this milling crowd of particles appeared a large, plain-faced clock with clear black figures and a second hand which moved

◖ NEWTON'S LAWS OF MOTION ◗

Isaac Newton stated three laws that describe the way that things, such as apples or planets, will move. Though these laws were set out back in the seventeenth century, people have still found no good reason to dispute them.

First Law: A body continues in a state of uniform motion if no force acts on it. This means that if you leave it alone it just carries on in the same inertial frame.

Second Law: The *rate* at which the velocity changes is proportional to the force.

This law is usually written as $F = ma$. Here F represents the force and m the mass of the object, while the acceleration a is the rate of change of velocity.

(In Special Relativity this law states that the force is equal to the rate of change of momentum, p, but at low velocities $p = mv$, so if the mass m is constant it comes to the same thing.)

Third Law: For every action there is an equal and opposite reaction. This means that if you push something it pushes back just as hard.

---

around in a series of perceptible jerks. Upon each advance of the clock the positions of the particles changed, and it could be seen that where two particles had a particularly strong line of force between them their movements curved in toward one another. As the particles moved, from one tick of the clock to the next, so the strengths of the forces between them changed with their new separations and so, in consequence, did the changes that each wrought upon the previous motion of the other. The motion was complex and ever-changing, but it could be seen that at each stage the new motion followed inescapably from the actions of the forces and that the forces were determined absolutely by the previous positions of the particles. With many particles involved it was complex, but the effects of each pair of particles followed a simple pattern.

"This process may be extended indefinitely. If you know enough about the movement of all the particles in the present, then you can predict where they will be and how they will move at any time in the future."

"No, I couldn't!" said Scrooge.

"Well, perhaps *you* can't; the calculation would not be easy to perform, I must admit. Perhaps no one is capable of making the calculation, but the information is there and so you might argue that the future of these particles is fully contained in their present. Newton's second law works equally well in reverse. If you know the forces, then this tells you how the motions of the particles are changing and this in turn tells you not only what they *will be* in the immediate future but also what they *were* in the immediate past. That sequence may be extended ever farther into the past, and the entire *history* of the group of particles is completely contained in their present.

"If you accept this Newtonian view, and Newton's laws are difficult to dispute, then these particles have no *independent* past, present, or future. For them time is undoubtedly just a coordinate which labels different states of motion and position but brings in nothing new. The whole history of this set of particles is completely predictable, utterly determined. The past of these particles is totally known, their future is fixed."

The Spirit paused and gestured widely to include both land and sky around them. "But what is the universe other than a very large collection of particles?" he asked rhetorically. "Everything in the universe is made up of atoms, and through gravity if by no other means they must all be affecting one another to a greater or lesser degree. When you consider the vast extent of Creation, it is clear that no one would have the skill and patience to make the necessary calculations; but given complete information about the state of all particles in the present it should be possible, at least in principle, to calculate their future state, should it not? If enough were known about the present you could, still in principle, calculate future behavior for all time to come. If such a calculation *could* be done, if the information is there which would *allow* it to be done, then the future history of the entire universe is determined, with no possibility of surprise."

"Can that be?" exclaimed Scrooge. "Surely the essence of the future is that it is, ultimately, unknown."

"I have described to you the view of the *deterministic universe*, a view that was widely held around the end of the nineteenth century and beyond. Many people who are not scientists think that is what all scientists do still believe. In such a universe there are truly no surprises, and nothing is really left to chance. There is nothing new under the sun, under *any* sun in fact. All that *shall be* is already fixed within the frame of what is *now*."

"You say that such absolute determinism might come if you had full knowledge of all particles in the present. Have you not been telling me, at some length, that there is no such thing as the present, that it depends on the viewpoint of the observer?"

"That is perfectly true, but it does not matter. In such an interlocked deterministic universe as I describe it does not matter who observes the events which we choose to call the present: since if all events are fixed, then one set is as good a starting point as any other. In this view the whole cosmos is a great machine, extending throughout all of space and time, with inconceivable numbers of interlocking gears, all turning together and acting upon one another in predestined fashion."

As the Phantom spoke the world around them seemed to change, and where there had been ground and sky, rock and tree, now Scrooge could see nothing but a vast array of interlocking gearwheels. Some were immense, and planets spun upon their turning. Some were tiny, far too small for Scrooge's normal vision to have shown them to him. On these turned the electrons that circled around individual atoms. Gears ranged over all possible sizes and were everywhere that he looked. In the rocks and the trees, extending far into the sky and deep within the ground in a seemingly endless chain of cause and effect.

"So what is time? What role does it play in such a fixed mechanical universe?" asked Scrooge's companion. Since he was the personification of Time, this was without doubt another rhetorical question and Scrooge treated it as such and allowed the Spirit to continue. "I have said that time is synonymous with change, but in such a deterministic universe there is no *real* change, only different slices through the frozen structure of eternity. What is the present? How, if at all, does it differ from the past or the future? The space-time structure of Special Relativity says that the present is not unique, it depends on the observer. Einstein and others from among your classical physicists would say that the present is an illusion, that time is a dimension not unlike those of space, and that each person has their time-line, their own path between the cradle and the grave, and that no point upon it is any more significant than any other. There is no absolute NOW any more than there is an absolute HERE. If time is but a direction, a line on which are threaded the successive events of your life, then where is the distinction between past and future? Newton's law can start from any point which you may choose to call the present and tell you the past or the future equally well, so where lies the difference? It may seem self-evident to you, but from whence does it arise?

"Consider that straight road of which we spoke before. If you take some place as your starting point, your HERE, and look at points along the road, then some will lie on your left and some on your right. Turn to face the other way, and those points that were on your left are now on your right. Can you so turn around that past and future will change places in this way? What is to distinguish them?"

Scrooge was by this time aboil with indignation. "Why the differences are obvious! You can remember the past but not the future. You can change the future, but the past is unalterable."

"If the universe is completely determined by the undeviating law of cause and effect, then certainly you cannot change the past but neither can you change the future. You have no options and no free will in the matter. As for what you can or cannot remember, is that the business of the universe?"

Scrooge was more incensed than ever by this remark. "Humbug!" he cried, "nothing but the veriest humbug! I cannot accept such illogical nonsense. I know that time exists, that the future and the past are not the same, and you cannot convince me otherwise."

The Spirit gave him a quizzical look. "Oh dear me! How very heated you have become! I would not be so quick to say that *that* is illogical nonsense. The notion of the deterministic universe may not be *true*, but it is fairly logical as far as it goes. If you really want to hear illogical nonsense you should wait until you meet my younger brother, who will come after I have left you. He will tell you things which are undoubted nonsense, though also demonstrably true. The frozen, deterministic universe which I have described was a view of Classical Physics. I felt it my duty to my elder brothers and sisters to tell you about it, but confident predictions are not always fulfilled, and the frozen universe has thawed considerably.

"You are right to feel that your deepest perceptions of the world should not lightly be set aside simply because someone tells you otherwise. The past and the future and the passage of time itself are so directly apparent that you do well not to dismiss them. You must be careful, however, that you do not confuse such unarguable knowledge of experience with mere habitual prejudice. You may do well to question the statement that past and future are no different, but you would not have the same justification if you chose to maintain that there must exist an absolute present, that there must self-evidently be absolute time, a universal NOW. Of *this* you have no direct experience. You can have had no personal experience whatsoever of simultaneous events at widely separated positions. You have never *been* in widely separated positions at the same time, a fact for which you should be grateful. You are always wherever *you* are and no place else. If you were to insist that time must be absolute everywhere it could only be through habit and prejudice. And you would be wrong."

"Do you then say that the idea of a fixed, deterministic universe is *not* true?" asked Scrooge, feeling much relieved at this.

"Time is a strange thing," answered his companion rather obliquely. "What is certain is that time plays a very special part in the nature of things. The idea of the deterministic and essentially timeless universe came from an extension of the predictable behavior of two particles to the behavior of many, and then on to the behavior of the entire universe. This is rather a grand extrapolation.

"Look around you!" commanded the Apparition. He pointed up toward the sky, which darkened straightaway to night, and Scrooge looked upward into the darkness of space. Indeed, as he attempted to look around him he found that the landscape upon which he stood had vanished, and he appeared to be afloat in space, looking down upon a planet that orbited around its sun. His time sense had again been slowed to such a degree that he was able to watch the planet for many revolutions in its path around the sun. He saw that on every occasion it retraced the same orbit, precisely and without deviation.

"There you see two objects, moving under the effect of their mutual gravity. For such a system the motion is indeed predictable. You could say where the planet will be far, far into the future, as far as you like. You can equally well say where it was at any time in the far past. Two objects that move solely under the control of the force between them are totally predictable, and their motion might be calculated for all time. For them the future would indeed be utterly determined. It is not even particularly difficult to make the calculation.

"This is an example of a regular periodic motion. One which repeats again and again. Many motions are periodic. Some, like the mutual orbiting of sun and planet or the swing of a perfect pendulum, are truly periodic and repeat exactly, without any difference from one instance to the next. Some motions are approximately periodic, like the sequence of the seasons. Each year you have spring, summer, autumn, and winter; but each year the weather will be different from the year before. There are many cycles in nature, and most of them have fluctuations from one cycle to the next.

"If you believe in a completely deterministic universe, then you may expect a grand cycle above all. A cycle that includes each and every other cycle in the cosmos. To those who were actually part of it, a deterministic universe might not look very different from one with an unpredictable future. True, complete and exact knowledge of the position and velocity of every particle in the universe might allow the future to be calculated, but you would not have that knowledge. The motion of individual atoms would still be utterly beyond your knowledge, as was assumed by the Shadow of Entropy. The ideas of statistical physics and increasing entropy, which depend on an inability to distinguish between the huge number of microstates available to the world, would still look just the same to you. Thermodynamics would be unchanged, and you would expect the Heat Death to come.

"However, if you wait long enough in what would seem to be the final Heat Death, then you would expect that eventually the motion of all atoms might be the same as at some time in the past; it does not matter which. It might be a long time indeed for this to come to pass, but surely in infinite time anything will happen sooner or later. If everything is then moving *exactly* as it was once before, then the future that would follow must be the same as that which *did* follow after the previous occasion because the present completely contains the future. Thereafter the future of the universe would develop exactly as it did the last time and everything, *everything*, would happen again. This would include all of human history and everyone would live again, and again, and again.... This grand cycle is known as a *Poincare recurrence*. In a deterministic 'clockwork' universe, which was in equilibrium and had infinite time available, it might be expected to happen. Not only do you not get any choice in what you do, but you have to do it over and over again!"

Scrooge was horrified at this vision. Not only did the idea of a deterministic universe seem to rob life of any purpose, but now it seemed that everyone would be sentenced to repeat this mindless dance at intervals, over and over again, time without end. It was the Wheel of Life, on a grander scale than he had ever envisaged, but also so much the less, so utterly futile.

The Spirit continued to speak, in what seemed a quite inappropriately cheerful tone. "You have seen that such determinism is correct, at least for a universe that contains only two objects. For three objects, of course, you would expect that the calculation would become more difficult. For four objects it would obviously be more difficult still and for a large number, say all the various planets and small asteroids within your solar system, it would be that much worse. For how many objects, all exerting forces on one another, do you think it is possible to calculate exactly their positions for all times far into the future?"

"I cannot say," replied Scrooge. "You had suggested that it should be possible in principle for the entire universe, though you seem to have backtracked from that statement. I believe it is possible in practice to calculate the future positions in space for all the planets and most of the larger asteroids, so demonstrably the number must be at least fifty. Yes, I should say well over fifty."

"The answer to my question is *two*," said the Spirit serenely.

"Two!" exclaimed Scrooge, moved once again to indignant protest. "It cannot be two! I know it is possible to predict the future positions of the planets very accurately, so the number must obviously be greater than that."

"No, I am afraid you are wrong. It is true that the positions of planets may be calculated quite accurately for many years into the future, but the calculations are only approximate. Accurate, but still approximate. There is a big difference in principle between a calculation which is completely exact and one that is approximate. An exact result will continue to be correct forever after. An approximate result may give a value that is almost correct but not exactly. More to the point, any small error may build up with time, and sooner or later the prediction will be totally wrong.

"If you had only one rock which was isolated in space, then the calculation of its future behavior would be very simple. In fact, it would not be a calculation at all, as the object would just continue to move in a straight line or to remain at rest, depending on which frame of reference you use to describe it. Two objects, like a sun orbited by a single planet, are more complicated, but in reality you have only one motion to calculate. Because momentum is conserved, any motion of one must be balanced by the motion of the other. There is only one independent motion, there are no irregular influences, and the orbit is completely regular and unchanging, as you have seen."

"If it is easy to calculate the movement of two bodies, then three cannot be all that much more difficult to deal with, can they? What about the planets in the Solar System?" persisted Scrooge, who found it hard to accept that just three objects could generate infinite complexity.

"I am afraid that three bodies are too many, and in fact their common motion cannot be solved exactly. When you predict the motion of *all* the planets, of which as you say there are many more than three, then your calculation may be very good and the planets turn up quite precisely where you predict they should. The problem is far too complicated to solve exactly, but you can get quite good approximate solutions for a long time ahead. There are a number of reasons why the approximation should be good.

"Because the Sun is very heavy it completely dominates the system. Each planet is attracted to the Sun far, far more strongly than it is attracted to the other planets. You can get quite a good prediction for each planet by assuming that all the others are not there at all and calculating the motion for each planet as if it were moving around the Sun quite alone. The effect which the gravitational force of one planet has upon another is not large, but it is there and will cause the orbits to change slightly as time goes on. The effect of the other planets will depend on their relative positions, and usually they are a long way apart. Sometimes the effect will be to slow one planet down a little. At another time, if the relative positions are different, it may speed up. Because all of the planets take different times to move around the Sun their relative positions will be ever changing. Such effects cannot accurately be included in your prediction, but as the forces are small it may take some time to build up any serious deviation from the simple calculation.

"You say that the positions of planets may be calculated years ahead, but of course a year is not a long time in the measure of the Solar System's motion. The earth takes that long to travel but once around the Sun. Some planets take longer, some less; but within your lifetime the errors in an approximate calculation may remain very small. However, they will build up. Slowly at first, but unpredictably the actual motion will deviate from your prediction. Wait for a sufficiently long time and present-day predictions will be totally incorrect. See what can happen in the long term with but three bodies. We shall look at three objects of similar size, so that no one dominates."

The space before Scrooge's eyes was lit by three glowing spheres. He realized that he was looking not at a binary star system but at a system of three stars, all moving around one another. With no obvious center the motion was complex and varied. Now the stars would all be well separated, now one would pass close to another and long fiery prominences would be drawn from its surface. Sometimes two would pass close by and perform a mutual pirouette while their third companion swung wide out to one side. Then all would move around, and the pairing would change. Eventually one of the stars swung closer than ever to a fiery companion. It executed a tight ballroom swing around this partner and soared off in a new direction. So different was this direction from its previous path that it was hurled away from its associates, and its light was eventually lost in the surrounding blackness of space. Its two remaining fellows now orbited serenely, one about the other, and settled down to a comfortably predictable motion of conjugal monotony.

"As you see, there are problems with any number larger than two. We may show this on a less celestial scale. Let us examine something which I know you have looked at many times before, a pendulum."

Once again the pendulum of Scrooge's clock appeared hanging in space before him, swinging to and fro and measuring out the seconds with a predictability that seemed to equal that of the orbiting planets themselves.

"You see the pendulum before you, but what you see at any moment does not tell you the whole story. A snapshot of a pendulum does not in any way capture its essence, which lies in its motion. It is in the interplay of position and movement that you have the essential nature of a pendulum. A picture might show the pendulum bob hanging below the point of suspension or at some distance to the right or to the left. It would not show how the pendulum is moving. It would not show that at the extreme of its swing it is at rest, that at its lowest point it is moving most quickly. You need a picture that shows not only its position but its movement as well. You need a portrait in *phase space*."

"And what, may I ask, is phase space?" asked Scrooge, assuming that he might. "You have told me already of four-space. Is this something similar?"

"In a way it is. It is a method of expanding your view of the universe by means of a picture, some of whose proportions may be measured along a range that is not one of the normal dimensions of space. In four-space one axis of the plot was along the direction of time. Now we adopt a different direction and draw the dimension, not of time but of momentum. Position and momentum. These together define the motion of a body, and it is position and momentum that we use to draw the motion of a simple pendulum."

The figure of Time reached into a large wallet which he had upon his back and began to rummage around. Scrooge heard him muttering under his breath "This satchel *will* keep filling up with entropy, with the waste heat of the World, not to mention *alms for oblivion*. I do not know how I can be expected to find things when I need them." Eventually he discovered what he wanted and drew out a large tablet or slate, together with a pencil.

"Right!" he said decisively. "Let us sketch out our diagram. Where shall we begin? The pendulum swings from left to right and back again, so we need use only one direction to plot its *position*. That leaves the other direction, the one measured up the page, to indicate how *fast* the pendulum is moving. First we draw a large cross in the center of the page. This gives our reference axes, crossing at the *origin* in the middle of the page. These are the lines from which we will measure position or velocity. To the right or to the left of the vertical line in the center of the page we measure position. Above or below the horizontal line we measure the momentum of the pendulum bob—above the line when it moves in a direction from right to left, below the line when it moves the other way.

"Let us begin with the bob pulled out toward the right. In this position it is at rest. It has a position away from the center but it has no motion, so we mark it to the right of the origin and upon the horizontal axis. When the pendulum falls from this position its image upon the paper will move toward the vertical line in the center, but its speed will increase and as its distance from the center drops to zero, so its speed and consequently its momentum will rise to a maximum and its semblance upon the paper will rise above the horizontal axis. At the center of its swing it moves with greatest speed. After that its position will move farther out toward the left and its speed fall again toward zero, so that at its leftmost extremity it is again at rest and its position marked on the horizontal axis.

"Now it will swing again toward the right and trace a similar curve, save that now it is moving in the opposite direction and so the line it traces will fall below the horizontal axis, until it is again at its right-hand extreme. At that point the cycle begins once more, and it retraces the same curve. See you the curve that I have drawn, which shows in one drawing the whole repeating history of the pendulum."

Scrooge looked at the drawing upon the pad and saw it was a circle. On either side were the occasions when the bob was far from the vertical but at rest. Top and bottom were the instances when the pendulum was vertical but moving with its greatest speed. Round and round the circle it would go, tracing out perpetual repetition.

"You see that the path in phase space is a circle, or rather, since I may choose at will the relative lengths at which I draw distance and momentum, the path is an ellipse. Of course you realize that this is not precisely the path which any pendulum will actually follow."

"Why ever not?" demanded Scrooge.

"Why, because there is always friction and other *dissipative forces* present. You have heard already how a pendulum, in common with all other systems in motion, will lose some of its energy on each cycle. As it loses energy to waste heat, so the swinging will decrease, each swing being a little smaller than the one before until the motion finally stops."

The surface of the tablet which the Apparition held in his wizened hand changed from a dull white to a clear glowing screen. The rough sketch of a circle on its front sharpened to a precise outline, drawn with computer-like precision. Scrooge could see that it was not precisely a circle but, rather, a very tight spiral which slowly wound in toward the center as the motion of the pendulum died away. Eventually the line had shrunk to a single point in the center, with no displacement and no motion remaining.

"There you see the final fate of any oscillation, however far you pull the pendulum aside to start it. The farther out you pull it, the larger the amplitude of the swing that will follow. But be it large or small, the energy will bleed away and the final state will be as here, with the pendulum hanging motionless below its pivot. It is the only possible outcome, however the process starts. You know how the motion will end up whatever the initial conditions. Sooner or later the graph of the motion is drawn into this central point. The point is known as an *attractor*; it is the only final outcome possible. Sometimes there may be more than one final state, and the outcome is no longer so clear. Consider this little toy, which may be familiar to you."

The aged Spirit held out his hand and on his wrinkled palm Scrooge could see a little pendulum that he did indeed recognize. It was an "executive toy" which normally sat upon the desk in his own office. There was a pendulum so mounted that it could swing in any direction. The pendulum bob was made of iron, and within the gadget's base were concealed three magnets which attracted the bob as it passed close to them. That was all there was, but it produced a bewildering sequence of movement, fully as random-seeming as the motion of the three suns which Scrooge had recently been shown.

"For this device there are not one, but three, final positions in which the pendulum might be found when it has lost its energy of motion. The bob may end up above any one of the three magnets and be held in that position by the magnetic attraction. Which magnet will be chosen? Obviously this depends on the earlier movement of the bob. Twist and turn though it may along its convoluted path, the motion is governed by Newton's laws as rigorously as the motion of the planets. The forces exerted by the different magnets depend upon the position and motion of the pendulum, and from these its future position and motion may be predicted, step by step. It should be a completely *deterministic* system. Where the pendulum will end depends on its initial motions; indeed it will depend on the position from which it is released.

"We may illustrate the dependence with this tablet. Each possible starting position may be shown in a different color, red if the motion that starts there finally comes to rest near one particular magnet, blue for another magnet, green for the third." The surface of the glowing tablet in the Spirit's hand was divided into areas of red, blue, and green to indicate the predictable final results of motions that began from each of the positions.

"You may see that within certain regions you may release the pendulum from any nearby point, and it will end up above the same magnet. What happens at the boundary between these regions? Is there an abrupt transition from one to another?" The view upon the surface of the tablet zoomed to show a region fragmented into areas of red, blue, and green. The magnification grew, and it could be seen that these regions were in turn finely divided into smaller regions in which red, green, and blue were still mixed together. However closely they looked, there was still a mixture of different possible results. However precisely the pendulum was positioned here, it might still end up opposite any one of the three magnets.

"The final state of the pendulum is completely determined by Newton's laws and by its initial position. Its future is *deterministic*, but the dependence is *fractal*."

"And what, may I ask, is a fractal?" inquired Scrooge coldly. He did not feel the last remark to be particularly helpful.

"I can best answer that by asking a question of my own. Can you tell me the length of the British coastline?"

"How does that answer my question?" replied Scrooge in some confusion. "I have no idea how long it is. Somewhere over a thousand miles for sure. I expect you can find the answer in an encyclopedia if you really want to know."

"I think not," the Apparition replied calmly. "The answer should not be given in any encyclopedia because there *is* no answer. How would you determine the length? You might take a map of the island and measure round the coast with a pair of dividers. That would give a result. You might set the tips of your dividers much closer together and repeat your measurement. Now you would step around

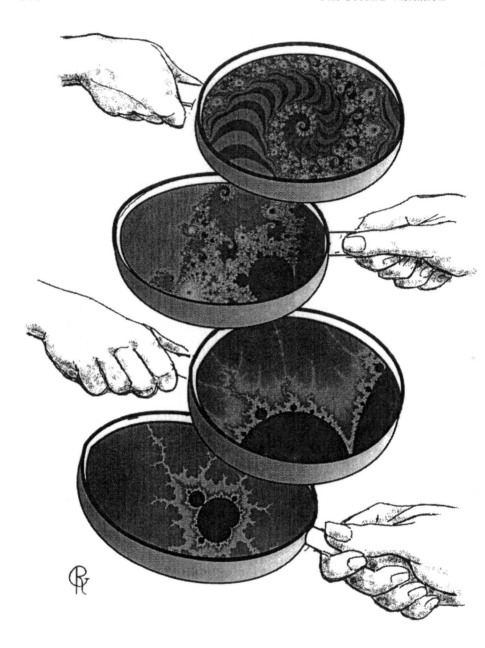

the outlines of bays and promontories which you had spanned across before, and the result would be different. If it were only slightly different that would not be surprising. You would say that a more accurate method of measuring would be expected to give a more accurate value for the result. The problem is that the result would be *totally* different. The distance around a bay could be three or four times as great as the width across it.

"The result you get depends on the scale at which you examine it. What is the correct measure? The value measured on a map with coarse dividers, with fine dividers, or the value you would get by walking round the coast and measuring with a ruler? If you did this, should you measure around the outside of every stone upon the shoreline? Should you measure in and out of every tiny hairline crack upon the surface of each stone? Should you measure around each atom in every stone? The answers would be quite different in each case. The coastline is a *fractal* object. It shows more and more detail the more closely you examine it.

"Nature tends to produce objects with a fractal form. Smooth spheres and cylinders are not common in nature; always you have finer and finer detail as you look more closely. You can create fractal objects on computers by making a sequence of repeated calculations, and one such sequence has achieved popular commercial fame. You can buy copies of the patterns it generates, featured on artistic posters. This is the Mandelbrot Set, by means of which a simple computer program can in principle generate infinite complexity."

Time's tablet lit up to show a strange black shape surrounded by an intricate pattern, full of fine curling flourishes. He produced a magnifying glass and focused on one section. This revealed further intricate structure, feathery and flower-like. A further magnification showed more delicate detail, and among the detail could be seen instances of the same strange black shape. Examination of these showed that they contained around them the same detail as the original. Sharper and sharper, higher magnification served only to show further detail, with no apparent end to it.

"There is no end to it," stated the Spirit. "The degree of detail is literally infinite. *That* is also the case for the plot of starting positions for your executive toy. The final state may indeed be determined completely by the initial conditions, but so critically that it is nonetheless impossible to make any prediction. However accurately you know the positions and velocities to begin with, there are still many outcomes that could result. These are not just slightly different futures but are totally divergent. That is what makes a mockery of the idea of classical determinism. A slight difference of conditions in the present does not make the future *slightly* different but can have enormous consequences. No matter *how* accurately you know the present, it can never be accurate *enough*.

"Tiny effects can have major consequences. This is often called the *butterfly effect*. The argument goes that the way the earth's weather changes is so critically dependent on earlier conditions that the flapping of a butterfly's wings in Borneo may later cause a hurricane in America. The hurricane will of course be 'caused' by many other things as well, but the effect of the distant butterfly might be just sufficient to decide between two possible paths of evolution for the weather, one of which will some days later include the onset of a hurricane in America while the other does not. Such a critical dependence on tiny effects shows how unpredictable the weather is.

"There you have the paradox of determinism. Yes, the future *may* be a direct logical consequence of conditions in the present and follow from them by an unbroken chain of cause and effect, but even so it may depend so critically on those precise conditions that, no matter *how* well you may know them, it is *never well enough*. So you see, even if the future is absolutely determined by the present, it is nonetheless impossible to have accurate enough information to make predictions. This in turn means that the future is *not* predictable."

CHAPTER 8

Never Mind the Destination, Enjoy the Scenery

Scrooge looked at the ancient figure of Father Time in some dismay. He was now looking incredibly ancient, more of a mummy than a man. To the casual observer he seemed as good as dead, even though he had not the decency to lie down and accept the fact. He appeared to shrink and wither still further even as Scrooge regarded him.

The Spirit nonetheless managed to rouse himself sufficiently to speak once more, in a voice so faint that Scrooge must needs strain

to hear it. "You might ask whether, if the future is indeed as random and unpredictable as the Shadow suggested to you, it means that it holds nothing *but* the promise of the Heat Death? Time may have come back as a player in the universe after it had seemed to be totally banished in a deterministic picture, but has it brought nothing but a steady descent to increasing entropy, decay, and utter formlessness?"

"Well, does it? Tell me, if you still have the strength," responded Scrooge.

"The final equilibrium of the Heat Death *may* be the end of hope," quavered the thin, tired voice, "but we are by no means there yet, and there may be new twists upon the route. What we *can* say is that the road is by no means a uniform decline away from all that is striking and interesting and toward something gray and tedious beyond all conceiving."

As he spoke the Spirit's voice seemed to strengthen again, and he stood more erect, while his wrinkled skin smoothed out to show an increasing flush of life. He still seemed old, no question about that, but now restored to a full and active vitality. "It is true that entropy increases in the world as we see it. Heat always flows from a hotter body to a cooler, and whenever it does there is an increase in entropy, in disorder. There is another step along the road to equilibrium. The source of available energy loses some of its previous store, the final morass of waste heat captures a little more of the energy that was available in the universe. That is true, but it is to look only at the two extremes of the situation. What happens to the middle—the systems that lie along the path from heat source to heat sink? This is the region where you yourself exist. Along this middle way the path that winds toward drab equilibrium is anything but uniform, and CREATION lies upon the road!"

The Spirit now stood tall and dynamic, towering above Scrooge with flashing eyes and flowing locks like some portrayal of an Old Testament prophet caught in the middle of a denunciation. "That is your position, *in the middle!* The energy that drives your world comes from the Sun. Slowly and steadily the Sun is running down and losing its store of available energy, but it has a long way to go. It will last your time and that of all your descendants, to the last syllable of recorded time. Some of this flow of energy passes through the earth and is eventually lost as low grade heat, mostly being radiated into the depths of space. Neither extreme of the sequence will normally concern you much. Your business is with the section in the middle, where the Sun's energy powers the machine of the earth's weather and the needs of your civilization. It drives the growth of plants and provides the flow of energy needed for animal life.

"You live your life very far from equilibrium, and for most of your life you show little sign of approaching it. The *Second Law of Thermodynamics* is not mocked in this. Entropy does indeed increase around you all the time. As you and other humans develop from conception and grow to maturity you become more complex and ordered structures. It does not appear as if your entropy increases in this, and indeed it does not. You do not become more disordered as you develop, but the *Second Law* gives no reason why you should. The rule is that the overall entropy of an *isolated* system should increase.

"The universe is, presumably, an isolated system, but its entropy *is* increased by your living. The overall entropy increase given by the heat and waste energy that you dump back into the world more than compensates for the local fall in disorder caused by your living and growing. You yourself are not an isolated system and, as I have said, very far from being in thermodynamic equilibrium. Energy flows through you all the time, from food, from sunlight, from breathing. You must have all of this to sustain your life, your local rebellion against a steady decay to equilibrium. If any animal is isolated and in consequence does not partake of this flow of energy, then its internal entropy increases as the *Second Law* requires. With no input of available energy through food to eat and air to breath, then entropy will increase and order and structure will decay. The technical name for this condition is *being dead.*"

"So you tell me that equilibrium may not quickly be achieved. Does this achieve anything other than to postpone the inevitable for a time?"

"Oh yes, it does indeed 'achieve something.' It achieves *everything*. When systems are far from equilibrium, with a generous flow of energy passing through, then interesting processes may happen. Creation may take place. What happens depends on how the different components of the world interact with one another. If nothing exists but isolated atoms and molecules, each doing its own thing with no reference to its fellows, then all states and conditions are of equal likelihood and worth. This was the assumption made by your previous guide, the Shadow; and in such case nothing will develop but random fluctuations. The road to equilibrium is then clear and direct. That is not how things are. In practice you have long-range interactions, and you have feedback. These can work together to create order, to build something new that was not there before."

"What exactly do you mean by long-range forces, and indeed by feedback for that matter?" Scrooge asked, promptly on cue.

"Long-range forces? Those are simply such things as gravity and electrical interactions. Nothing terribly new and remarkable there, you might say. The existence of such long range interactions means that particles have an effect upon one another most of the time and not just when they happen to collide, as is assumed in a simple kinetic theory of gases and other materials. Each particle may be influenced by others afar off, and this reduces their selfish independence. The behavior of one may affect others, and this encourages coherent action. It is not just every particle for itself, each doing its own thing; but now they may all work together. One illustration of

⅏ Universal Gravity ⅏

The first example of a long-range force appeared in Newton's theory of universal gravity. He postulated that all bodies in space attracted one another, however far apart they might be, with a force that depended on their masses and varied as the inverse of the square of their distance of separation.

Two objects with masses m_1 and m_2, separated by a distance r, will attract one another with a force

$$F = G\frac{m_1m_2}{r^2}$$

where G is the *constant of gravitational attraction*.

Newton was able to show that this law was sufficient to explain the observed motions of all the planets (to the precision at which they were then known; corrections given by Einstein's theory of relativity are measurable for the orbit of Mercury).

what may result is given by the formation of stars, stars such as your Sun, which is the source of the available energy used by all life upon the earth."

As had happened before Scrooge found that, with little fore-warning, he was once again apparently floating in space. There was no sun or planet before him this time, and he understood that no star had yet been formed nearby. He was aware, though just how he was not really sure, that before and around him drifted a great cloud of gas, mostly hydrogen, and that from this material a star would be born in due course.

Within the cloud the density of gas fluctuated as the molecules moved around at random. At one moment the density at one position would be slightly greater than at another, but as the molecules moved these effects would average out and the regions of greater or lesser density would fluctuate. Scrooge sensed that at one particular position the density, for whatever reason, was quite significantly higher than average. If there had been no interaction between the molecules, then the thermal motion would soon have smeared out this local concentration, as it had removed lesser fluctuations; but there *was* interaction. Every molecule pulled at every other with the weak tug of gravity. Molecules in the surrounding gas were attracted slightly more strongly to that region where already there were more molecules to pull at them, and on average yet more were drawn in. More were drawn in than escaped to the regions of lower density, and so the concentration of the gas increased. As the concentration further increased, so did the pull of its gravity upon other gas molecules in the regions around.

The initial core of higher density grew. It grew tentatively at first and with the ever present possibility that random movements of the molecules would wipe out all the gain won by the first faint tugs of gravity. As the concentration in the central region grew, however, so the imbalance of the forces felt by neighboring molecules became ever greater, and they were pulled more decisively from the low density regions into the growing heart of the future star. As surrounding regions became depleted by this steady loss of gas, so the gravity exerted by their lower density was progressively less able to hold back the accelerating drift away from them. To the region that already *had*, more was given. The regions that *had not* lost even that which they had. All of the gas of atoms and molecules round about was swept into the growing star.

As more and more matter fell toward the center of this new star, its mass increased. As a consequence, the potential energy released by matter falling into its gravity became steadily that much greater. This energy appeared as heat, and the star became ever hotter as it contracted and drew more of the surrounding gas into itself. The star's temperature rose to a white heat and above, and it began to glow brightly. Its center rose to temperatures much higher even than the glowing photosphere near its surface from which came most of the visible light. The temperature in the heart of the star eventually became so high, the energy with which the molecules collided so great, that they were able to produce nuclear reactions, as can the particles in an earth-bound accelerator. The center of the star became a nuclear furnace and released yet more energy as heat and light. The outward blast of light and other radiation from the hot star pushed away the infalling matter, and *radiation pressure* prevented further collapse.

Uncomfortably close, a star now burned in front of Scrooge, close twin to the familiar Sun he knew. The star had attained a form of temporary stability. It was no longer collapsing or drawing in new material because the outward pressure of radiation from the nuclear fires within held these processes at bay. As long as these nuclear reactions continued, it remained virtually unchanging, constant in size, and pouring out floods of heat and light. This all came from the nuclear energy that had previously been locked inside the matter from which the star had been formed.

The pause in the star's collapse was temporary; gravity was merely biding its time and would have its final triumph. Sooner or later the star would have burned all of its supply of nuclear fuel, the nuclear fires would die, and there would be no more radiation pressure to prevent its collapse. From that moment the inexorable shrinking would begin again. This temporary respite could be of quite long duration, however. The sun of Scrooge's own solar system was but a little way through the pause in its own collapse.

"There you saw one of the more obvious effects of long-range forces." The Phantom's voice broke in upon Scrooge's vision of the fiery star, which thereupon faded from before his eyes. "You have seen gravity produce structure on a broad scale. More importantly it has produced a localized source of available energy, a star much like that Sun which is the origin of life and light upon your own world. The structure produced was large and crude. There was no sign of the intricacy, the fine convoluted tracery that is the mark of life. For that we look to the effect of feedback, of nonlinearity."

"And what may nonlinearity be?" asked Scrooge. "For that matter, what is *linearity?*"

"Linearity in physical interactions is the first reluctant step taken by theory as it moves beyond the untenable assumption that particles do not interact with one another at all. All interactions are held to be between two, and only two, consenting particles and to be in no way the concern of any others. If there are many particles present, then each may interact with every other one, but the way in which it interacts with any one is in no way affected by the presence of the others. It is no longer every particle for itself, but it is still every *pair* of particles by itself. The potential caused by the presence of many particles is just the sum of the potentials caused by each one separately. There are assumed to be no collective effects, where the presence of other particles can enhance the effect of each. Above all, there is no feedback."

"If I might ask the obvious question," said Scrooge as if on cue, "what do you mean by feedback? Is it like my responding to your remarks by asking questions?"

"You could, I suppose, say that your questions provide some sort of feedback from my statements, but they lack one vital aspect of feedback as it affects physical systems. Feedback occurs when the result of something is included among its *causes*. The chain of cause and effect becomes twisted and circular and this can, as you might imagine, have significant consequences. Because feedback affects the sequence of causality, of cause and effect, you can see its consequences in logical arguments as well as in physical systems. Consider these two signposts." Two posts appeared before them, securely embedded in nothing in particular. One carried a sign that read:

> **Many hands make light work.**

The other notice read:

> **Disregard the first notice, it is not true.**

"These two notices give no logical problems when taken as a pair. You might debate whether the first is true, but the existence of both notices presents no problem either way. One may be true and the other false. Which is the case depends on outside information, on external circumstances. The situation is altered if we change the first notice." The second notice still read as before, but now the first was altered to read:

> ## The second notice is untrue.

"Now you do have feedback. The result is that the interpretation of each notice affects the input to the other, the *input* on which its interpretation is based. Taken together, their interpretation forms a closed loop. There is still no particular difficulty with these notices. If the first is deemed to be true, then the second is false, which is of course consistent with the first being true. This interpretation is stable. So, however, is the interpretation that takes the first notice to be false and the second true. This also is consistent. The remarkable feature is that you have two interpretations, each equally valid but mutually exclusive. The interesting feature is that you can choose either interpretation, *without any change of external circumstances*. There is a two-valued instability. The situation may be even more extreme if we look at a notice that provides its own input." Both notices faded from sight, to be replaced by a single one, which read quite simply:

> ## This notice is not true!

"There is a case where no reasonable interpretation is possible. The notice looks like a fair statement, little different from the previous ones, but its meaning is unstable. Is it true? If so it is false. If it is false, it is true. Whereas in the previous example each notice could be true or equally it could be false, in this case it can be neither. This strange and uncomfortable behavior is because the notices are self-referential; there is feedback. Feedback can give equally strange results in physical systems. It is usually seen when the state of a system helps to determine what happens to it next. Feedback can serve to make a situation more calm and stable, as when you use a thermostat to control your domestic heating. A thermostat gives negative feedback. If your room is too hot, the feedback reduces the heating; if too cool it turns it up. The net effect is to keep the temperature roughly constant. Positive feedback can have quite the reverse effect and produce dramatic effects apparently from nothing. You can see, or rather hear, this by placing a microphone too close to a loud speaker."

The Apparition held up a microphone clasped in one scrawny hand. Behind him could be seen the towering speakers of a powerful sound system, though anything less like a pop star than the elderly figure of Father Time would be hard to imagine. He moved purposefully toward the nearest speaker, holding the microphone tentatively in front of him as if he expected it to explode. In a sense that was what it did. Scrooge clapped his hands over his abused ears, as there was an unearthly, deafening screech, totally unlike any sound that had been evident before. It rose in intensity to a level that jarred him to the very bone, and then he could hear it no longer. He took his hands away, realizing that the stack of speakers had vanished and that the Spirit was in the middle of addressing him again.

". . . so you have seen that a dynamical system that is linear and without any dissipation, such as a friction-free pendulum, will move

forever in a completely predictable fashion. Such behavior is monotonous and it is also quite unrealistic. Nothing in normal experience is completely without dissipation of some sort. Even the planets in their orbits through the heavens lose some of their energy by tidal effects and by collisions with the small amount of material that lies in their paths. When energy is lost by dissipation and not replaced, then motion will eventually come to a halt. Whatever the amplitude of that motion it will decay steadily, if perhaps rather gradually, toward a final limit point: an attractor. In the case of a pendulum, this attractor is simply the state in which all motion has ceased and the pendulum bob is hanging straight down, but other instances may show more complex behavior. This is particularly so when energy is fed into the system at the same time as it is lost to dissipation.

"This is the condition that we must examine. This results in the far-from-equilibrium activity that takes place upon the earth. Energy is fed into the system. Ultimately it comes from the Sun, from which it cascades down through a great chain of processes. Energy is lost from the system. Dissipative effects of one sort or another will steal energy and lose it to the final morass of low-grade heat. Where there is energy input then the system may go on running indefinitely without running down. Energy is transferred from the energy source to the final sink, and in the process entropy increases. The bit in the middle, the part of the process that is of most interest to you, need have no part in this increase of entropy. If the system is linear and without feedback, then the behavior of the final attractor will be smooth and regular: a limit cycle. The motion would be smooth and regular forever. It would not change, and neither would its entropy change, either increase or decrease.

"If there is significant feedback, then things can become complex. Feedback and self-regulation are striking aspects of living organisms, either individually or as species. A simple example of feedback is shown by the way in which a population changes from year to year."

The formless background that had surrounded Scrooge and the Spirit changed, like the blurred image of a projected film when it abruptly sharpens into focus. They were standing in a large formal garden, in a clearing surrounded by shrubs and bushes. At one end of the clearing was a statue of Pan, playing on his pipes at the edge of a large fish pond. In the pond Scrooge could see the dark, secret shapes of many fish.

"Within that pond there is to be found a population of fish. Each year they breed, and the following year there will be a new population. If there were no feedback, if each fish could live quite independently of its fellows and each produce several offspring, then their number would grow steadily and exponentially. Of course fish do not live forever, and each year some would die; but provided the birth rate could more than keep pace with the deaths, the population would still rise. However, the fish are not unaffected by their companions; they must all live in the same pond.

"In the pond the supply of food is limited, and this puts a limit on the population. If there are few fish, then the population may grow because there will be enough food for all. If there are far too many fish, then they may eat up most of the food and practically all will die, so the population will fall dramatically. You might expect the population to stabilize at some steady value, a limit value, such that sufficient new fish would be born each year to balance the number that died, and that this population would thereafter remain constant year upon year as long as the food supply remained unchanged. Whatever the population initially you would expect it to approach this optimum number, possibly oscillating around it slightly to begin with but eventually settling down to a constant value.

"The example shows many of the features of nonequilibrium systems. There is an annual breeding, an input of new fish that parallels the input of available energy from the Sun to the earth. There is the loss of fish, from old age or starvation. There is feedback, because the rate at which fish die depends on the number there are competing for the same food. If the fish breed slowly and the *growth parameter*, the increase in the number from year to year, is small, then the population does indeed rise and stabilize at an optimum value."

The Phantom stretched out his hand and behold a line was drawn upon the surface of the pond, as if upon a blackboard. This curved to begin with but then held steady down the center of the pool, to indicate the steady population year after year.

"As the breeding rate increases and the build-up of the population becomes more rapid, then the behavior changes. At some critical value of the growth rate for the population the number over successive years abruptly begins to oscillate. One year there is a relatively large number, and they starve themselves; the next year the number is much smaller and can increase, so that the next year the population is back to the higher value. Year after year the population swings between these two numbers, never settling down to one."

The Spirit spread the fingers of his hand, and the line marked upon the waters split abruptly into two—one marked the larger population, one the smaller and the value in successive years swung from one to the other.

"If the growth rate for the population increases further, then abruptly you see another doubling of the possibilities; now the populations go through a four-year cycle, with different values in each year between. A further increase in growth rate and the cycle of population doubles yet again, repeating every eight years. Eventually there comes a critical value for the growth parameter above which there is no recognizable period, no repetition of the population. The fluctuation in the number from year to year is apparently quite random, and the fish population has become chaotic." The plot marked on the surface of the pond split again and again, bifurcations coming more and more frequently as the growth rate rose until there was nothing but a blur, with no discernible pattern in it.

"A real pond full of real fish is of course a more complicated thing. The fish may suffer from some disease. There may be a swarm of insects which land on the pond and provide more food for the fish. All manner of things could cause the population to vary and complicate the situation, but this unpredictable behavior does not rely on any such complication. The population growth in an ideal pond may be given by a mathematical model of great simplicity, one that takes the number of fish to increase by a constant factor each year but also to fall by an amount that depends on the previous year's population. Even a model as simple as this will show all of the oscillations and the final chaotic behavior I have described. Simple processes can produce very complicated behavior. That is the message of chaos.

"Chaos is *not* the same as randomness. Random behavior is completely unpredictable, but it is also without any sign of order or pattern. Chaos may be unpredictable, but it can generate patterns, often very detailed and intricate patterns. Consider once again a pendulum, since this is such a simple device. This time we will feed energy into the pendulum each cycle and allow it to be lost by dissipation."

A post appeared between the pond and the statue of Pan. It supported a jutting bracket from which hung a pendulum whose bob was moving round and round in a circle. A mechanism at the point of suspension gave the bob a small kick every time it went round, and the energy so gained was lost to friction and air resistance during the rest of the cycle. As a consequence the pendulum bob circled round and round, repeating the same motion over and over again. Or was it the same motion? It certainly looked much the same.

"There you have a simple system, a pendulum that goes round in a circle again and again. The motion will not run down because it is being fed with energy to make up for the loss through dissipation, as the earth is fed by energy from the Sun. The motion repeats, but it does not repeat exactly. There is an element of chaos, as you may now see."

Where the moving pendulum bob passed through the air in its motion, it now left behind it a fine colored track, like the colored vapor trail left by aircraft in some aeronautical displays. Round and round went this thin line, tracing out successive revolutions of the heavy bob. As he watched closely, Scrooge could see that the orbits were not repeated exactly. Each was similar to the one before, but it was not identical. The tracks left behind were not *quite* the same. As more and more fine lines accumulated from successive orbits, the whole began to resemble a tangled skein of thread with successive turns looping and twisting through one another without apparent order. The motion was visibly chaotic and indeed it looked to Scrooge to be entirely random. He said as much to the personification of Time.

"As you now see it, the record of the motion may appear random and formless, but it is not. There is structure here even if you do not see it." The Apparition reached within his pouch and drew out what looked like a long pole. He pressed upon some concealed catch and a long, wicked blade swung out at right angles to the haft. Scrooge realized that he was seeing a *flick-scythe*. The Spirit swung this dangerous looking implement twice, almost faster than Scrooge could follow, and excised a thin slice from the middle of the loop. He held this out for Scrooge to examine, grasped firmly in a thin but powerful hand. Over the area of this cross-sectional slice Scrooge saw a scattering of fine dots, which he realized marked the points where each orbit had crossed

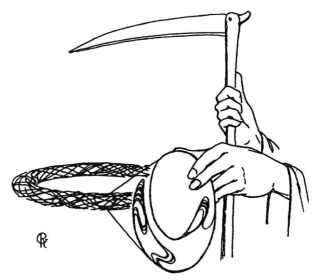

that area. Scrooge was intrigued to see that, even as he watched, more and more points were added to the slice. He realized that, despite the fact that it had been cut out, it was still continuing to record the points at which the circling pendulum bob crossed the region. There seemed no pattern in the sequence of these crossings. A new dot would spring up at one point, the next at some position quite far removed. Although the *sequence* of the crossings showed no clear pattern, their *positions* most certainly did. Across the slice which the Phantom held in his hand spread a series of thin looping lines, curling and twisting and splitting and forming a most intricate tracery. The lines were made visible by the sequence of dots which lay along them. Initially there were many gaps as the number of revolutions of the pendulum and the number of dots which showed its points of crossing were limited. As time went on and more and more dots appeared, the pattern became sharper and clearer as every dot fell only upon the lines; and the more dots there were the more clearly the lines could be seen.

"There you see structure, fine detail born from chaos! This pendulum is a simple system, if not quite linear; but when there is dissipation combined with a source of energy to keep the motion from dying away, then you find chaos, and within the chaos you can find order.

⟪ᴄ Chaos ᴐ⟫

Chaotic systems are those where, due in general to nonlinear effects, infinitesimal changes in the starting conditions may soon produce totally different behavior. Chaos may be seen in processes which are described by very simple equations, and such behavior may easily be produced by computer programs. Most of the research into chaotic behavior has been done with computers, as you can then known what is producing the chaotic behavior you see.

An example of chaotic behavior is given by the way that smoke will rise smoothly from a cigarette and then abruptly develops irregular swirls and eddies.

A feature of chaos is that although the process seems totally unpredictable it may often display strikingly ordered and regular behavior. This appears to have little dependence on the initial conditions, and the same cycle of behavior, or *attractor*, may result whatever the initial conditions.

"In many systems quite simple rules and mechanisms combine with a driving source of energy and with dissipation to bleed the energy randomly away. This may produce chaos, and from the chaos order may arise. Often the structures created may be intricate and delicate in the extreme, showing detail of almost fractal complexity. This may be seen in purely chemical processes in which the behavior depends on chemicals already present, so introducing feedback. Such reactions may show regular patterns in space and regular pulsing in time. They are not alive, but such behavior resembles a simple living organism.

"Self-organization is a feature of complex systems, complex in the sense that they include many atoms. To see such pattern and structure you must have dissipation of energy and nonlinear behavior, but when you have both of those, then you may see intricate forms appear that have little to do with the details of the underlying mechanics at the atomic level. Simple time-reversible behavior on the atomic scale *can* result in complicated and creative behavior on the large scale. You get newly created forms that were not in any way built into the system by some precise initial plan. They do not arise because some program was written in at the start of time which would produce them. It is more remarkable, more creative than that. The forms are new. The behavior of such systems is so critically dependent on initial conditions that they are quite unpredictable, and when you get such self-organization into intricate patterns then something completely new has come into the world.

"You may see this behavior in a familiar example." As the Spirit spoke, white flakes began to fall from the sky and to settle on the grass. It was snowing. The flakes grew steadily larger, as often happens when it is snowing, but this time they continued to grow past any reasonable expectation. Scrooge realized that it was not that the fall of snow was becoming heavier, but that the flakes were appearing larger to him. He was seeing normal snowflakes with ever increasing magnification, and now he could see them falling as distinct lacy shapes. Each flake was a six-pointed star, delicate and full of fantastic detail. Each was a six-pointed star, but apart from that no two were the same. He was surrounded by a seemingly infinite variety of shapes and details that shared no common feature except that they all had sixfold symmetry and of course they were all pretty cold!

"There you see an example of detailed structure that has arisen in the process of freezing water from a cold cloud. Apart from a slight bias toward the sixfold symmetry, no feature of the individ-

ual flakes was built into the initial conditions. The growth of the
flake has a degree of feedback built in, because any region that grows
will have a greater surface area, and so more ice will be able to freeze
onto it. From that comes all the variety of structure and detail that
you see. The considerable differences which you can observe even
between neighboring flakes have arisen because no two will have
traced *quite* the same path through the clouds, and their development depends so critically on conditions that the flakes end up completely different from one another."

"All of this growth of variety which you have shown me is undoubtedly impressive," said Scrooge. "But from what you have told
me it is still ultimately doomed. Perhaps in the short or the middle
term there may be creation, but is it not fated to come to nothing
in the final equilibrium of the Heat Death?"

"When there is final equilibrium throughout the universe, then
indeed you will have the Heat Death, but that has not yet come
about. Nor is there any sign that it is imminent. The universe is locally quite inhomogeneous, with huge temperature differences between stars and their surroundings. On the large scale also the universe is not static, for it is continuously expanding. The stars and
distant galaxies are moving ever apart, the more distant receding
more quickly that those nearer to hand. The whole universe is expanding, opening out like a star-shell in a great cosmic fireworks
display. "This may continue for ever, or it may be that the mutual
attraction of gravity will slow the expansion to a halt. Perhaps from
such a halt the continuing pull of gravity will reverse the motion and
the universe will fall in upon itself, contracting down to a smaller
and yet smaller size. Eventually the universe, which was first born
in the Big Bang, may complete the cycle and shrink down to a final
Big Crunch."

Scrooge once again found himself apparently drifting in the
depths of space, looking upon the vast panoply of galaxies that
stretched away upon all sides and that he could see as moving ever
apart. He realized that his sense of time had again been distorted,
and by amounts greater even than on that occasion when the Shadow
had sent him forward to experience the Heat Death of the universe.
Every cosmic component was moving apart, but was he right in
thinking that the separation was slowing down. He could not say,
as he had no way of knowing if his sense of time was at all reliable;
but after some unknown further time he felt that the expansion had
ceased, that the more remote stars were no longer receding. Indeed
they now appeared to be coming together, however slowly.

The contraction continued until all the stars in the sky seemed to be tumbling together, everything in the whole of the universe was compressing down again to a single tiny space. As the matter of the universe collapsed under gravity and became more compressed, so its temperature rose, repeating on a grand scale what happened in the formation of stars. Scrooge fell with the rest, miraculously protected from the extreme conditions while everything fell in toward the Big Crunch. Abruptly the final moment of compression came and Scrooge felt not so much a crunch as a flop, as if he had collided with a large interior sprung mattress. He looked around and found that was just what he had done: he was in his own bed, in his own room, and it was still night.

✵

The Third Visitation

In which Scrooge has his last visit from a Spirit, but one whose message is stranger and, apparently, more foolish than all the others.

Scrooge hears of the generous all-encompassing dreams of Nature. How anything that can happen in some sense does happen, how everything that might be is in some sense really present, with no possibilities excluded, until an observation is made. It is the observer or the measurement, some such interaction with a system, that creates a unique reality.

He learns of interference, of waves, particles, and uncertainty. He learns of fluctuations and of particles which come from nowhere and return again. He learns of correlations and connections that seem to disobey some of the laws of physics. But all of this, he learns also, is reliably verified by experiment.

CHAPTER 9

The Ghost in the Atom

Awakening in the middle of a prodigiously tough snore and sitting up in bed to get his thoughts together, Scrooge had no occasion to be told that he was once again in his own bed chamber. It was still well before dawn, and his room was dark, though the darkness was not quite as intense as when he awaited the arrival of the previous Spirit.

His room was now full of shadows and false shapes, suggesting to the eye all manner of monsters and phantoms. The patches of darkness came and went with arbitrary dispatch as his eyes attempted to pierce the surrounding gloom. As he became gradually more accustomed to the dim light he noted that one patch of shadow was more definite than the others. He fixed his eyes upon this shape and beheld a solemn phantom, draped and hooded, coming, like a mist upon the ground, toward his bed. He started up and looked wildly around. To his surprise and dismay he saw that many other shadows had assumed this same hooded form. Around his bed they crowded, many replicas of this same uncertain shape. Some were the merest wisps of being, vague suggestions of a shape. Some were more substantial and loomed darkly about his bed.

In the very center of the group stood one most definite and nearest to an absolute solidity. It was shrouded in a deep black garment that concealed its head, its face, its form and left nothing of it visible save one outstretched hand. But for this it would have been the more difficult to detach this figure from the night and separate it from the darkness by which it was surrounded. Scrooge felt these shadows to be tall and stately as they came beside him, and their mysterious presence filled him with a solemn dread. He knew no more, for the shape before him neither spoke nor moved.

"Am I in the presence of the last of the Spirits, whose arrival was promised to me?" asked Scrooge.

The shade answered not, but pointed onward with its hand.

"What manner of things will you show me?" Scrooge pursued. "Are you about to show me shadows of the things that have not happened but will happen in the time before us? Am I now to follow you?" He imagined that the upper portion of the garment was contracted for an instant in its folds, as if the spirit had inclined its head. He could not be sure, and that was the only answer he received.

Although well used to ghostly company by this time, Scrooge feared the silent shape so much that his legs trembled beneath him, and he found that he could hardly stand when he left his bed and prepared to follow it. The phantom moved away, and Scrooge followed in the shadow of its dress, which bore him up, he thought, and carried him along.

They scarcely seemed to enter the city, for the city rather seemed to spring up about them and encompass them of its own act. It was not the city as it might be expected in the small hours of the night but more as it might appear in early evening, full of the bustle of prosperous activity. They stood before the doorway of a luxurious hotel, up whose imposing steps surged wave after wave of well dressed men and women, all of whom possessed the indefinable aura of very large amounts of money.

They moved closer, and the spirit's hand was pointed toward a large though tasteful notice, prominently displayed beside the door. Scrooge saw that this announced the occasion to be the annual dinner of an association of financiers, the guest of honor being one LORD SCROOGE, life president of the association and director of a long list of companies. These were listed in full, but they are hardly of interest to the reader, though you may be sure that Scrooge read each and every one of them with the greatest interest imaginable.

****

The ghost conveyed him until they reached an iron gate. He paused to look around before entering. A churchyard, walled in by houses, overrun by grass and weeds, the growth of vegetation's death not life, choked up with too much burying. The spirit stood among the graves and pointed down to one, newly dug but meaner and less distinguished even than the other graves in this forgotten place.

"Before I draw near to that stone to which you point," said Scrooge, "answer me one question. Are these the shadows of things that Will be, or are they shadows of things that May be, only?" Still the ghost pointed downward to the grave by which it stood. Scrooge crept toward it, trembling as he went, and, following the finger, read upon the stone of the neglected grave his own name, SCROOGE.

****

The ghost conducted him through streets familiar to his feet. They stopped by a door he knew as well as any other. It was his own place of business. There upon the nameplate by the door in silent proof stood the names of MARLEY and of SCROOGE. He turned to the ghost. "What is the meaning of these visions?" he asked. "There seems no order to them. They seem to me to be such scenes as cannot reasonably be considered together. Come Spirit, explain. Speak to me!"

****

****

"Aha, there you are. I spy you now!" The voice did not come from the shrouded shape, which vanished from his sight. Scrooge turned abruptly at being so accosted and saw a new figure, one as different from the brooding presence he had so far accompanied as it was possible for him to imagine. The newcomer was short and round of face. He had enormous, implausible feet and baggy trousers held up by brightly colored braces. He was bald, with a bushy fringe of surprising green hair and a large, round nose of vivid red above a wide mouth painted in a mischievous grin. He was, in short, without any possibility of mistaking, a clown.

"Who are you?" asked Scrooge somewhat disdainfully. Though the shrouded figure of his recent guide had filled him with dread, he had at least had a presence more appropriate to serious discussion than this ludicrous new arrival.

"Why, I am your third Visitor, come to talk to you about the Quantum World. Were you not told to expect me? I looked for you in your room, but you had already left. Why didn't you stay where you were meant to be? It was purely by chance that I happened to observe you, but then of course it always is."

Scrooge gazed at him in amazement. This was certainly not his idea of a tutelary Spirit. "If you are indeed the final Spirit of whom I was forewarned, then who or what was the figure that I followed to this place?"

"Why no one in particular. A mere nameless amplitude which was present in your room and you became included in its superposition of states. It was fortunate that I observed you here, and so of course here is now where you are. I am the Spirit of Quantum Physics, come to reveal to you a quantum description of the world. A clown has a different viewpoint on reality from that of sober conventional people, and quantum physics has without doubt a different viewpoint. Quantum Mechanics has been called the *Ghost in the Atom*, but it would be more appropriate to speak of the *Clown in the Cosmos*, for the truth I shall tell you will seem too ludicrous and nonsensical for serious people. It is surely the message of a clown, but as is often the case the statements of the Clown are found to be true. I will show you a world that is the same world that you have always seen, but seen quite differently. It may appear so strange and distorted that you will be hard put to recognize it, but it is still the same world, the real world, the only one that you have."

"What do you mean by states and amplitudes?" asked Scrooge. He was still not convinced of the wisdom of speaking to this absurd figure, but if he was indeed the third Visitor whose arrival had been promised, then Scrooge thought that he should at least make an attempt to learn from him. "I had understood that a state is a word used to describe the condition of a physical system, but what do you mean by an amplitude?"

"An amplitude gives a possibility, a choice for the system, one of the parts which goes to make up the totality of experience. Come and I shall illustrate this."

"Come? Come where?"

"Why, to the Circus of course!" The Clown cavorted off down the road, and Scrooge hastened to follow him. They rushed through the dark streets, and as they went Scrooge became aware of the presence of a subtle, ghostly procession down the middle of the road. Dim and uncertain he fancied that he saw, flamboyantly parading, all the normal complement of a circus. There were jugglers and horseback riders, acrobats and trampolinists, performing bears and fire eaters. Leading the procession was a stately line of elephants richly draped, while snarling lions brought up the tail. Down a narrow alley they all marched, and Scrooge followed them until they came to an open space in which billowed the spreading canvas of the Big Top.

Scrooge and his companion passed within, the Clown insisting that they should enter by crawling beneath the canvas. As far as Scrooge could see this was from sheer perversity. Within was the circus ring, this one most unusual in that it was completely surrounded by tall curtains that quite obscured the audience's view. The Clown motioned Scrooge to sit beside him in the exact center of a large block of empty seats. No sooner were they seated than the curtains twitched aside to reveal a display of juggling. The performers were agile, but the show was in no way unusual as far as Scrooge could tell. For a brief period they watched, and then the curtains swept across again. A moment later they opened to reveal a trampoline act. Later they opened upon the sight of a lion-tamer making his beasts go through their tricks.

"What is the relevance of all this?" demanded Scrooge. "And indeed how do the various acts enter and leave the ring? I see only one present when the curtains open, but there seems no way out for any of them."

"There is no way for them to come and go, and they do not. The ring contains a superposition of amplitudes, one for each of the acts in this quantum circus. They are, in a sense, all mixed together upon the stage, but when we observe we see one or other of them."

"That does not sound a plausible explanation to me," retorted Scrooge. "All we actually see is that, whenever we look, one act or other is being staged. Is it not far more reasonable to assume that there is some route by which one act leaves, and another replaces it? That explanation would seem to fit our observations just as well and to be more sensible by far than to say that somehow they are all on stage together when we do *not* look. How can they be?"

"You can be sure that, in some fashion, they are indeed all present together because you see interference between the different amplitudes. You can get a mixing of the possibilities that are present together and one possibility affects another. Observe further!"

Scrooge watched as the curtain was swept aside again and again to reveal different acts. Some seemed normal enough, but then he watched a bareback riding exhibition in which the riders were *bears!* Every now and then a strange mixture of acts would be revealed. He marveled at the predicament of a lion tamer whose charges were not only obviously ferocious but compounded his problems by breathing fire. And as for the elephant's trapeze act, he would not easily forget it!

The Clown turned to Scrooge. "I am joking of course. Interference of amplitudes is not quite like this."

"I knew it couldn't be!" thought Scrooge in satisfaction.

"You may not find the real situation to be much more reasonable though," warned the Clown. "Come with me and let me show you." He led Scrooge down from the block of seats and around the side of the ring. As they walked Scrooge noted that the sawdust that covered the ground had imperceptibly given way to a fine yellow sand. Abruptly the curtains drew apart. Farther and farther apart. The curtains on either side receded into the distance to reveal a great expanse of sea, and he saw that they were walking along a beach beside it.

"The sea is calm tonight, the tide is full," remarked the Clown conversationally. "And on this bare beach the sand blows and is gone. Only, where the sand meets a transverse fence, see it spreads out through a fine crack, fanning in smooth profusion from that point." Indeed, there were a series of low fences running across the beach as one often sees. One after another, they were visible all along the beach, running across the sand and out into the water. In the one nearest to them Scrooge could see a narrow crack in the fence between two planks. Fine sand was being carried by the wind up against the fence and some drifted through this gap. On the opposite side of this crack the sand fanned out from the hole to form a broad, smooth mound.

Another fence a little further down the beach had two narrow cracks side by side. The mounds of drifted sand that had blown through these holes had overlapped to form one broad, smooth hump, centered halfway between them. There was nothing very remarkable about any of that, as far as Scrooge could see. "You will see nothing very remarkable about any of that," said the Clown in a prosaic manner. "The sand drifts through one gap or through the other, and grains of sand from both sources lie on top of one another to form a broad smooth mound. Much what you would expect really.

"But what are the wild waves saying?" he declaimed abruptly, apparently abandoning the topic. As far as Scrooge could tell they were not saying anything and indeed were not very wild, being the merest gentle ripples. He said as much. The ripples ran across the surface of the sea until they ran against one of the fences that stretched far out from the beach. There was a gap in this fence also and locally the ripples found their way through and spread out widely upon the farther side.

"The waves may not be dramatic, but see how they run wild and free as they escape from the constricting confines of the gap in the fence and spread out in the clear water beyond. What are they saying? They are saying 'up and down, up and down,' of course. That is all waves ever do or say. It is their nature to wave up and down and it is this same nature that causes interference when they meet."

They walked a little farther down the beach to the next fence. As Scrooge had by now anticipated, this one had two gaps in it, close by one another, and the waves squeezed through and spread out from each. As the gaps were so close together the spreading wavelets crossed one another in the process.

"There you see the meeting of the waves, and when they meet they must speak with one voice. Up or down, how should the surface of the water behave? The waves must fight it out between them. If both wish to go up together, or both to go down, then they may readily agree. At such position the water will go up and down more strongly than it would for either wave alone. They work together with a will, and this is called *constructive interference.* If, on the other hand, one wave is saying 'up' while the other says 'down,' they must agree to differ and the surface of the water gets no instruction and does not move at all. The efforts of the two waves cancel out completely and nothing happens. This is termed *destructive interference.* This is the nature of waves. In some places they work together, and the net disturbance, the net energy involved, is greater than the sum of what both would contribute at that point. At other positions they cancel, and the net effect is nothing."

"So you tell me that particles and waves are very different," said Scrooge. "What of that? I do not find that to be in any way surprising. It would be more amazing if you were to tell me they are the same!" he said jocularly.

"Aha, you've guessed, you've guessed!" cried the Clown in delight. "That is just what I *do* say. Come and see!"

Scrooge looked at their path ahead. The beach stretched onward, broken at intervals by the low fences across the sand. In the distance the sun was setting and stained the clouds with a dramatic red that spilled over to stain the surface of the sea. It was a glorious sight, reminding him of the advertisements for package holidays which he frequently saw on the television. Unconsciously he looked around for the outline of a television screen bordering the picture. He saw one! He watched astounded as they approached the giant frame that surrounded the wonderful picture, came up to it and passed beneath it. Ahead of them the beach scene vanished abruptly. They were in a strange place devoid of features and when he looked behind, he saw the beach stretch away toward the sunset *behind him!* He realized that he was seeing, from behind, a *picture* of the beach that was formed on the screen of some enormous television tube. Either the tube was enormous, or perhaps he was now small. From his recent experiences that was an option he could not rule out.

"Right!" exclaimed the Clown in a purposeful way. "Let's get to work and sort out these electrons." He took off his coat and hung it on some form of magnetic coil nearby; then he took an enormous wrench from a capacious trouser pocket and hit the coil with it. The picture of the beach scene shrank and dimmed to a patch of colorless light in the center of the screen. Looking at it closely Scrooge could see a sort of bright, grainy effect as little sparkles came and went on the inner face of the screen. The Clown told him these were the *scintillations:* the individual flashes electrons produced as they struck the phosphor screen.

"Now we need some sort of barrier," he muttered. He reached again into his pocket and drew out a dark object. It drew it out and out, to a length that surely could never have fitted within the pocket. When finally it came free it could be seen to be a long roll of some dark material, which he promptly unrolled to make a partition stretching across the width of the television tube. "Now I need some holes. I am sure I have a hole in my pocket somewhere." He searched industriously in both trouser pockets and came out with something which he slapped on the surface of the partition. Scrooge could see that there was indeed a hole there, as the Clown illustrated by putting one of his fingers through and wiggling it about. He then repeated exactly the same procedure, and there were two holes, side by side.

Throughout this demonstration Scrooge had not been looking at the television screen. When finally he glanced that way he saw that the uniform patch of light produced by the electrons had changed and that now light and dark bands ran across that region. In each band Scrooge could see the sparkling of individual electrons as they struck the screen, but there were far more in the bright bands than in the darker ones between.

"There you have it!" proclaimed the Clown. "There you see both wave and particle behavior at one and the same time. The shifting sparkle that comes from the detection of many electrons shows that we are seeing distinct particles. Each electron makes its own individual flash upon the screen, and the bright patches are the regions where there are more flashes. The very fact that there are bright and dark bands is a sign that there is interference. The bright bands are the regions of high activity, the dark bands the regions where there is destructive interference and nothing much happens. Interference

❧ WAVES AND PARTICLES

In quantum physics the concepts of particles and waves have become mixed together. Both of these words carry across fairly definite pictures from large scale experience. Enough of these carry over to the small scale for the names to be useful, but it must always be remembered that of things on a small scale behave in ways foreign to our intuition, which has been developed only from experience of large scale phenomena.

Waves spread over a considerable volume, and waves may combine to give interference, adding in some places, subtracting in others. Waves carry energy and momentum, but it is spread throughout the wave, with each part of the wave carrying a little.

We think of particles as little, hard, round things, but their essential properties are that they are found in one place or another and that all the energy and momentum they possess is found there. It is all or nothing with particles.

The quantum amplitude combines both wave and particle properties. Like waves it spreads over a wide area and *shows interference*. Interference is the principal experimental evidence for quantum physics.

When *observations* are made, all of the energy and momentum of the "particle" are found in one place, as you would expect for particles. *Observations* have a very important role in quantum physics, because any observation must affect the thing observed. You cannot *see* something without scattering light from it, and when *photons*, the particles of light, interact with that thing they can and do affect it.

---

is a property of waves, and so you see the electrons are both wave and particle!"

He sounded very satisfied as he said this, but Scrooge was not so sure. "I thought you said that interference was produced by waves coming from different places."

"Oh, certainly. The interference is produced by the two holes. See what happens if I close one!" He produced a large cork from his bounteous pocket and plugged one of the holes with it. The pattern now produced by the electrons upon the far screen was a diffuse spread of fluctuating sparkles, flickering points of light over a broad region, with no sign whatever of the bright and dark bands Scrooge had seen before. "The interference does require both holes, as you can see."

"Do you mean," began Scrooge as he tried to sort this out in his mind, "that in some way the electrons that pass through one of the holes will interfere with those that pass through the other, like the different waves that came through the two holes in the fence before?"

"In a way," replied the Clown, "but the answer is more singular than that. It is not electrons from one hole that interfere with electrons from the other. Each and every electron manages individually to interfere from both of the holes. Let me show you." The Clown removed the cork, then took the large wrench from his pocket again and marched over to a glowing filament at the far end of the television tube. This was the electron gun, which emitted the beam of electrons. He gave the device a hearty blow with his wrench, and the bright glow dulled considerably. "Now the rate at which electrons pass through is much less, as you can see on the screen." He pointed back toward the television screen and indeed the number of bright flashes was much reduced. The glowing patch was much fainter, and it took longer to build up any recognizable features. But when Scrooge had grown accustomed to the changed intensity he could see the same pattern of light and dark bands, exactly as he had seen before.

"Now the time between the emission of successive electrons is much greater than the time it takes them to travel from the source to the screen. There will only be one electron in flight at any one time, and yet, as you can see, there is still interference. It is not interference *between* electrons; indeed it is every electron for itself. Electrons, however, are very interfering bodies and manage to interfere with *themselves* without any assistance."

"That I do not understand at all. How can the fact there are two holes affect what an electron will do. It cannot go through both of them after all!"

"Can you be quite sure about that?" asked the clown with an exaggerated quizzical expression.

"Of course I can! That would clearly be nonsense. Ask anyone you like."

"Objection! Objection! I object Me Lud! That is merely hearsay evidence." Scrooge looked in surprise at the Clown and saw that he was now wearing a tiny barrister's wig perched on his bald pate. He was standing with his monstrous feet spread dramatically apart and his thumbs hooked into the lapels of a legal gown. The effect was slightly spoiled by the fact that his brightly colored suspenders were outside the gown.

"You say that an electron cannot go through both holes, but can you say that of your own certain knowledge. Have you ever looked?"

"Well no, of course not," began Scrooge, "but it stands to reason. . . ."

"I put it to you that you should make the effort forthwith. Let the witness examine the evidence!" With these words he whipped off his wig, rushed round to Scrooge's side, and handed him a large

magnifying glass. Feeling somewhat irritated by this clowning
but genuinely intrigued by
what he might see, Scrooge
moved closer to the two holes
in the partition and peered at
them through the lens. He was
unable to see anything and reported this.

"Well, what do you expect?
You cannot *see* something unless light has actually hit it and
interacted with it, you know.
The light here is rather dim,
and I expect that no portion of
it has actually hit any of the
electrons passing through, so of
course you will not be able to
see them. You cannot make an
observation without some interaction."

"But there is some light
here, even if it is dim. Surely a
little must hit each electron as
it passes through!"

The Clown looked severely
at Scrooge over the top of his

spectacles. It occurred to Scrooge that the Clown had not been wearing spectacles before and had apparently put them on solely so that
he could look severely over the top of them. "I do not know what
you mean by *a little* light. I hope you do not imagine that light can
be split up into arbitrarily tiny amounts just for your convenience!
You have just seen that electrons, which you believe to be particles,
will also behave just like waves. Does it not follow that light, which
you believe to be a wave, must also come as particles. That's only
fair you know!"

"I do not see any of that!" protested Scrooge hotly. "I have on
occasion heard people speak of light waves and the wavelength of
light, so I suppose light must be a sort of wave, but if it is it certainly does not follow that it should be a particle as well. I am not
at all sure I am convinced by what I have seen that electrons are

both particle and wave, but even if they are it does not follow that light must be the same."

"Well, it is!" snapped the Clown. "There is no doubt about it, the evidence is quite clear. Light behaves as a wave, but when you detect it or when it interacts in any way with something else it behaves as a particle. You must have seen television pictures taken in the dark using *image intensifiers*, to show field mice in their nighttime activities or something of that sort. Have you not noticed how the picture is rough and grainy, like a poorly developed old photograph, but with the bright grains appearing and disappearing randomly all the time? Each bright grain is caused by the detection of one *photon*, one particle of light. The situation is perfectly symmetrical. Electrons are particles, but they are also waves. Light travels as waves, but it is also particles. If there are not enough of these photons, or light particles, then it is likely that none of the photons will hit the electrons close to the holes, and so of course you will not see which hole the electron went through. No *observation* will have been made. You haven't much chance of seeing without light you know. Let me give you more light."

The Clown gestured with his hand. Scrooge could not imagine at whom this gesture was directed since they were quite alone, but immediately bright spotlights shone upon him from every angle, and he was bathed in brilliance. Looking again at the region of the two holes he could now see faint flashes of light scattered from passing electrons. Some were near one hole, some near the other. In no case did he see a flash near both of the holes. It was one hole or the other, every time.

"It is just as I expected!" he cried triumphantly. "Electrons do go through one hole or the other, as any reasonable person would expect."

"I wouldn't know," responded the Clown. "Not being a reasonable person myself, I do not know what is reasonable to expect. I know only what actually happens. You have now looked, as I suggested you should, and you have seen the electron actually pass through one hole or the other. Now I suggest you consider your verdict.... BUT LOOK THERE!" he cried out suddenly. Scrooge whirled around. He could not see anything different, only the featureless patch of light on the television screen, the featureless patch of light WITHOUT ANY INTERFERENCE BANDS! The interference had completely vanished.

"And that about sums it up, ladies and gentlemen of the jury."

The Clown adopted a declamatory manner and addressed himself to a nonexistent audience. "You have seen my client examine the two holes in the partition and observe the electrons as they come through, but once he has observed them you have seen that as a consequence of the very process of observing them, of the interaction between the light and the electrons, which is the vital ingredient of any observation, no sign of interference now remains. I put it to you that we have demonstrated beyond reasonable doubt that when an electron encounters two possible paths, in this case through two holes in the screen, it will avoid making a difficult choice and take *both* options. This leads to interference between the different paths, as long as there are no witnesses to the path taken, since any observation must require some *physical interaction* to record information about the electron's position.

"If an electron should be observed near one of the holes, then the court may know that where it is thus seen to be is where it is. The process of observing the electron requires an interaction with a photon that will affect the electron and so changes the system to exclude any other possibility. As the electron has then definitely not come through the other hole, there is of course no interference. An act of observation must of necessity involve some physical interaction that changes the system, and such observations are responsible for the loss of interference. Accordingly I would ask the Court to find the *Witness* guilty."

"Come now," exclaimed Scrooge. "Enough of this Tomfoolery! Are you seriously telling me that an electron can go through two holes as long as you do not look to see what it is doing? Are you further telling me that when you do look the electron will always go through one hole only, but that the consequences will then be quite different?"

"No, of course I am not *seriously* telling you anything! As a Clown I never do *anything* seriously. But you are right. Your summing up is perfectly correct; the observation is part of the process."

"But how can that possibly be?" protested Scrooge. "How can my observations have any effect? The electron cannot know whether I am watching it or not. My observing it can have no effect at all."

"Oh, but it can, indeed it must," responded the Clown. "You cannot observe anything without affecting it. Under no circumstances! Never! If you observe anything in any way, there must be some physical interaction with the thing observed, something must convey information about it. If you *see* something this can only happen if some light has scattered from the thing observed. Light consists of photons, which have energy. If a photon interacts with something it may well affect its amplitude. The observation is the process that extracts the information, not the use you make of it. The bright light would have destroyed the electron interference from the two holes even if you had not looked to see where they were. Whenever information becomes available about something, that is an observation even if you yourself do not choose to look at it.[1]

"As I said, the observation is part of the process. If you will come with me to the Show, I will show you what I mean."

"Show, what show?" asked Scrooge, bemused.

"Why to the Greatest Show in the universe: the universe itself! Come along!" Quickly the Clown rolled up the barrier he had placed across the TV tube and thrust it again into one of the pockets of his baggy trousers. He led Scrooge down into the far depths of the tube. As he passed the dimly glowing wire filament of the electron gun he paused to thump it with his wrench. The filament thereupon glowed as brightly as it had at first. He led Scrooge through a small door which they discovered behind the electron source. As he reached the door Scrooge glanced backward and saw that the picture of the beach scene was once again visible on the face of the tube, drawn with myriad tiny flashes as electrons were guided to the appropriate parts of the colored phosphor screen.

On the other side of the door they found themselves in a theater, a veritable old-fashioned theater. There were red plush seats rising in tiers from the stage. Above were galleries and boxes, adorned with carvings heavily gilt and glittering chandeliers hung pendant from the ceiling. Down the central aisle went the Clown in a series of cartwheels that ended as he vaulted upon the stage. Scrooge did not feel inclined to follow his example and slipped into one of the chairs in the empty, dusty theater.

The Spirit stood in the center of the stage. Behind him was a series of mirrors with gilt frames. The house lights dimmed, and in these mirrors various aspects of the natural world might be seen re-

---

[1]This may not be entirely true. At present, quantum mechanics has a problem known as the measurement problem, which makes it difficult to say at what point the amplitude does "reduce" to one possibility only. The difficulty is that if the light that interacts with your object is also a quantum system, then it should have an amplitude that is a superposition of all possible outcomes of the interaction, in which case nothing is really resolved. Some theoreticians say that the outcome becomes definite and unique only when it is observed by a *conscious mind*. Other theoreticians say this is nonsense. The problem is discussed in more detail in Chapter 5 of my companion book, *Alice in Quantumland.*

flected. Scrooge could recognize a swirling galaxy of stars in one, and in another he saw a rain-swept landscape in which a rainbow arched across the sky. In yet another was displayed a regular array which he believed must be the structure of a crystal. He saw an awesome fiery prominence rising from the surface of a sun. He saw the intricate mouthparts of a tiny insect, magnified so large as to fill the stage. He saw the diverse crystalline perfection of a group of snowflakes. A host of such scenes faded one into another, so that they were continually being replaced and formed a backdrop that illustrated all the variety of the physical world. It seemed that the mirrors successively illustrated every single thing that was to be found upon the world or in the universe around: Air, Fire, and Water; earth and Heaven. It gave a whole new meaning to the term multimedia.

The Clown began to speak, and the frames behind him cleared of their images, becoming simple mirrors that showed only his reflection. "Nature's picture of reality seems to differ from yours," began the Clown gently, clearly speaking to Scrooge. "You believe that everything can be discovered clearly and precisely if only you look sufficiently closely. Your picture would be drawn in detail, with a ruler and a sharp pencil, but Nature paints with a broader brush. Nature has rather a *dream* of what is and what may be. Nature's view of a particle is generous and ample and is usually called an *am-*

ꙮ Superimposition of Amplitudes ꙮ

The superposition of amplitudes is probably the most remarkable feature of quantum mechanics and is intimately tied in with the phenomenon of interference, which provided the dominant experimental evidence for the theory.

The central dogma is that you cannot describe a system in any more detail than is possible to observe experimentally, at least in principle. If there are various things that might happen, which are not ruled out by some form of available information and there is absolutely no way of telling which is the case, then there is an amplitude present for *all* of them. The overall amplitude is a superposition of *all* possibilities. If an electron might have gone through either of two holes, and there is no way of telling which, then automatically there is an amplitude present for it to go through each. To that extent you can say that electron *goes through both holes* because there is an *equally valid* amplitude for each. You may say that what is not forbidden is compulsory.

This might appear at first sight, and indeed at all subsequent sights, also, to be total nonsense. It is, however, the *only* picture that has been devised that is capable of allowing interference between different possibilities, and this interference really *does happen.*

◁ REDUCTION OF AMPLITUDES ▷

A system is described by an amplitude, which is in general a superposition of all the possible results which you might get for an observation. As noted below, an observation requires a physical interaction whose consequences differ for the different possibilities, and if you have not made such an observation none of these results is excluded. In that case the total amplitude includes each possibility.

When you do make the observation, however, you get one particular result. Beforehand it might be possible to predict only the probabilities of various results, but after the observation there is no doubt. The result was what you saw it to be. The amplitude for the system has changed *as a consequence of the observation*. It reduces to contain nothing but the amplitude for the result seen. If a second observation is made immediately after the first it will give the same result, as that is the only amplitude present. (This condition may not last for long, as the amplitudes do change with time.)

Observation is a term used to include any sort of physical interaction whatsoever that could distinguish between the states of a system. Such an observation will select one from among the results it can distinguish. There is no element of choice in this, what you see is quite random; but from the moment of the "observation" it becomes definite.

---

plitude. It encompasses all that a particle may do. Everything is allowed that is not, for some reason, forbidden. I may even say that everything is encouraged. A particle says 'Can I be here?' and Nature replies 'Yes, of course you may' and includes that option. Again the particle says 'May I be over there also?' and Nature adds in that option as well.

As the Spirit described this addition of all possible amplitudes, Scrooge saw that in the mirrors behind other small images of the Clown appeared, while his original reflection shrank a little with each new addition. "The total amplitude that describes any system may be a *superposition*, a sum of many possibilities. It may include very many, for Nature's Dream is complex. Everything affects everything else, if only to a very slight degree, and the Dream, the Amplitude, contains it all."

The mirrors behind now teemed with multiple copies of the Clown; some large, some small; some sharp and clear, some very dim, to see: the merest ghost of a Ghost. From all around Scrooge a cry rang out of many childish voices all crying "Look behind you! Look behind you!"

"There's nothing behind me," said the Clown.

"Oh *yes* there is!" came the chorus.

"Oh *no* there isn't," countered the Clown.

"Oh *yes* there is!"

The Clown turned abruptly to look behind him and behold: there was indeed nothing there but his own reflection, solitary before him. This image stuck out its tongue at him and then proceeded soberly to copy his every move. However, in the section of mirrors that were now behind the Clown's back, the array of multiple images began to build up once more.

The Spirit turned to face Scrooge. The many copies of himself behind the glass ran swiftly across to fill those mirrors which were now out of his sight. "There you see the role of the observer. There may be many amplitudes in the superposition which describes a system; but when you look, when an *observation* is made, then you see one thing or another and only one possibility remains; all the other possibilities vanish from the amplitude. The possibility that remains is now quite clearly the only one, since after all it has just been observed. This process is called the *reduction of the amplitudes.* An observer may think that he or she is "only looking," but that is not possible since the very act of *looking* requires that light must interact with the object examined. Such *looking* forces Nature's hand; a choice is made, and reality is changed. The Witness is responsible for the Act, the Reporter responsible for the real state of the World."

✿

Chapter 10

Virtually Un-certain

Scrooge had sat still for as long as he could bear, listening to the Clown's remarks. At last he could contain his impatience no longer. "I must say that all you tell me sounds very vague and nebulous!" he called out. "And it doesn't seem to have much to do with particles or waves either!" he added firmly.

"The time has come,"
the Clown remarked,
"to talk of many things,
Of waves and probability
and what the future brings,

With energy and particles
all lurking in the wings."

As he spoke the last word, the Clown ran into the wings of the stage but straightaway came back out again wearing a schoolmaster's gown and mortarboard. He was carrying a lectern that he set up at the center of the stage. Hanging on the front of it was a notice that read, "Lecture in progress."

"The second thing you must note," he began, "is that the amplitudes are by no means vague and nebulous! They are as precise and definite as any of the notions of classical mechanics."

"I find that hard to believe," called Scrooge. "And if that is the second thing I must note, what is the first?"

"I shall get round to the first thing eventually, perhaps, sooner or later," replied the Spirit. "But waves I shall speak of now, for the amplitude has the form of a wave."

"What sort of wave?" interrupted Scrooge, who had decided that he must keep asking questions if he was to get any clear information from this volatile Spirit.

"It is not such a wave as you will see upon any body of water in Nature. Rather, the amplitudes are ripples upon the sea of Nature's grand unconscious. There are many such waves, spreading out from the random casts of observation. Like ripples upon water, they may run abreast and pass through one another on their way. Like water waves, or waves of any form, they have their highs and lows, their ups and downs; and when the waves come together they may add or they may subtract. The waves will interfere; and as with any wave, this interference may be *constructive* or *destructive*. The total effect may be that much greater, or it may be *nothing*."

"All right!" called out Scrooge. "Perhaps these amplitudes are a sort of wave, but in that case where do particles come in? The two are still very different."

"Why, the particles come in when you observe them. When else could they come in? The amplitude is Nature's plan, her dream of how things could be, and it combines many possibilities. When Nature is observed, then you see particles with all their energy and momentum. Each particle you see is in one place only, in that place and no other. The connection is made through the intensity of the wave. An intensity is given when an amplitude is squared and its magnitude found. Previously when waves on water, or sound waves in air, were considered it was understood that this intensity told you how energy was distributed in the wave. It showed how much energy there was in one region, how much in another. Now we see that the energy is concentrated in the particle, wherever that may be observed. All of the energy is then in one place, the place where the particle is seen to be. There is no question then of *energy* being spread smoothly across the extent of these waves. What is distributed is *probability*. The intensity of the quantum wave gives a *probability distribution*."

"And what, may I ask, is a probability distribution?" demanded Scrooge, realizing just too late what the Clown was bound to say.

"It is a distribution of probability," answered the Spirit obligingly. "As I have just told you, it does not give the amount of en-

---

⟪ Probability Waves and Probability Distributions ⟫

In the small scale quantum world you cannot justifiably speak of what a particle is doing when you are not looking at it. You might have prejudices as to what it should or could be doing, but if you do not examine it there is no way, even in principle, that you can *know*. You can only say what is actually observed when you (or someone or something else) looks and *observes* its condition. A central feature of quantum physics is that you cannot predict exactly what such an observation will show for any one particle, you can only calculate relative probabilities for different results. This behavior seems strange, and to many people it seems quite unsatisfactory, but it is very strongly supported by experimental evidence. The best description of "what the particle is doing" is given by an *amplitude*, or *wave function* that contains the best information about the particle. Like all waves this may be positive or negative in different places, and the ways may interfere. They may either add, to give a greater value, or subtract, to cancel one another out.

If the wave function is multiplied by itself, thus producing a value that is always positive, you get the *probability distribution*. The magnitude at any point gives the probability that a measurement will discover the particle to be there. The definite prediction you might expect from classical physics has been replaced by a set of probabilities. It is *more likely* that the particle will be observed in some places than others. It is possible for the probability to be very large in one region and virtually zero elsewhere, and when you are dealing with large numbers of particles this can give quite precise predictions for the relative numbers found in different regions.

---

ergy at different places; instead, it gives for each of those different positions the probability that you will observe the particle, with all its energy, to be at that position.

"Where the probability is high, then that is where you are most likely to observe the particle. Where the probability is low, you are unlikely to observe the particle. When the probability is zero, then you have no chance. In that case you can say definitely that you will not find the particle there, but that is the only completely definite statement. Otherwise the particle could be seen anywhere."

"What is the use of that?" protested Scrooge. "If you may observe the particle to be anywhere when you come to look, then your amplitude is meaningless, it tells you *nothing!*"

"Ah, but that is not so! For one particle the wave function makes a poor prediction; but when you have many particles, all described by the same amplitude, then you will find that more are observed wherever the probability is high, fewer where it low. If you have a very large number, then the probability distribution will accurately describe the way the particles are seen to be distributed. As the particles are distributed, so also will be their energy. A classical light wave is just the probability distribution for the photons, the particles of light, when you have a very large number of photons present. When you look at things on a large scale, you may in one case think you are dealing with a wave and in another case with a particle, like a dust mote. Both of these cases are aspects of the quantum wave function in situations where you have a *very large number* of fundamental particles involved."

"I am sorry, but I cannot accept any of this. It is quite against common sense," said Scrooge, though to tell the truth he really didn't feel sorry at all.

The Clown upon the stage grew still, and he seemed to grow. He did not so much grow in any way larger but, despite his ludicrous appearance, he seemed to grow more *significant*, more **REAL**. Scrooge was abruptly aware that, although he had been feeling sufficiently assured to ask carping questions from his position in the audience, he was nonetheless in the presence of a Being, a Spirit, a personification of the basic structure of the whole universe. One who, as such, was *more awful* than the shrouded figure who had so unnerved him at first.

"MAN, YOU TAKE TOO MUCH UPON YOURSELF! It is not your place to say what you will or will not accept from Nature. She is as She is, and you have no choice but to accept Her. Your *common sense* has little relevance here, for it is but the distillation

of your experience of the com-
monplace, of happenings and
circumstances similar to those
you have encountered before.
You have seen how things work
out in such cases and so to them
you may apply your common
sense. But not here. You have
had no direct experience of the
quantum world, so here your
common sense is worthless.
"You have **no right** to judge
what Nature may or may not
do. No one asked you! Nature
does what Nature wishes, and

all you can do is accept reality. This is the real world, however
strange and even nonsensical it may seem to you."

"Sorry!" muttered Scrooge, considerably daunted by this outburst "But it does seem completely nonsensical."

"Of course it does!" cried the Clown more amiably, with a little skip in the air. "Of course it seems nonsensical. If it did not,
would I be here to explain it? It *is* nonsensical to your way of thinking. But it is also true! "The wave function," he continued, in tones
as calm and level as if he had never been interrupted, "is precisely
related to the energy and momentum of the particles, and it develops in a way that is as regular and predictable as the motion of any
planet bound by Newton's Laws. The proportions are exact, the relations invariable. The energy of a particle is strictly proportional to
the frequency of the wave that describes it. Frequency is the rate at
which the wave moves up and down, the number of times the amplitude changes from a trough to a peak and back again in one second. The wavelength, the distance between successive peaks or successive troughs of the wave, is similarly related to the momentum of
the particle. As the momentum becomes greater, so the wavelength
decreases, in strict proportion. The actual size is given by a quantity, called *Planck's constant*. This is very small, and so the energy
and momentum of any individual particle is very small, very small
indeed. On your coarse scale of being, you would be quite unaware
of anything so small; and it is only when you come upon many,
many particles together that you are even aware of their existence."

"It seems I must accept my rebuke and also all that you are
telling me," said Scrooge, still rather grudgingly. "You are saying in

✺ Wave and Particle Properties ✺

A wave is described by its *frequency and wavelength*.
A particle by its *energy* and *momentum*.
At the quantum level we see that waves and particles are the same thing, so these properties must be related. The connection is given by the *Einstein Relation*:

$$E = h\upsilon$$

where E is the energy of the particle, υ is the frequency of the wave, and h is a constant called *Planck's constant*.
and the *deBroglie Relation*:

$$p = \frac{h}{\lambda}$$

where p is the particle momentum, λ is the wavelength of the wave, and h is again Planck's constant.

effect that anything can happen, and you cannot tell what will happen until it has been observed and then all is revealed."

"No, no, no," cried the Clown, covering his bald head with a thick bushy wig, which he produced from his pocket and from which he proceeded to tear out handfuls of hair. "No, no and no! Or, to put it another way, NO! You cannot have been listening! It is just not true that *anything* can happen. Only those options for which a probability is present may be observed, and when you have large numbers of particles, as you so often do, the probabilities predict very accurately what you will find.

"You have even managed to get the end of your remark wrong as well," he added more calmly. "When you have an observation, then *something* becomes definite and is revealed, but not all, not *all*. One measurement cannot measure everything, and there are things which can-

not be measured precisely at the same time. When a measurement is made it will give a result and immediately afterward the amplitude for the system will be the one which corresponds to that result and that alone. Measure the same thing immediately after, and you will get the same answer as you did before because you have the amplitude that corresponds to that result, that and no other. You will have, for the moment at least, a *definite value* for that quantity, whatever it may be. But you cannot make measurements of all quantities at one and the same time.

"Consider momentum," he remarked, leaning confidingly on his lectern. "The momentum of a particle is given, accurately and precisely, by the wavelength of its associated wave function. The momentum may be predicted with complete confidence if the wave function has but a single wavelength, like this." Across the mirror behind him was drawn a wiggling green line. Up and down, up and down, in complete uniformity and with precise separation of its successive peaks the wave spread across the mirror behind the Clown. With no change in its appearance the wavy line spread across each one of the mirrors on either side. Even these were unable to contain it, and the line extended beyond the last of the line of mirrors and continued on either side through the air of the theater, vanishing into the distance with no variation of its regular form.

"That is a wave with but a single wavelength. It describes a state for which the momentum is exactly known; but as you can see, such a wave is exactly the same in all places. There is no change, no increase in probability for one region as against another. It just goes on and on the same. The position at which you might observe the particle is totally unpredictable. For this wave there is no uncertainty in the value that you would find for the momentum, but the uncertainty in the position is infinite.

"In most cases you will have some idea where the particle is. You may have detected it, and you will then know, at least approximately, where it was. Should you detect it again straightaway you can predict approximately where it must be, since it would not have had time to get very far away. Its wave function must be *localized*. There must be a great probability for finding the particle in that vicinity. But how can we alter our wave to give this localization? You have seen that a single wave with a single wavelength shows no change, no matter how far afield you look; it is completely *unlocalized*. The solution is to add together many waves and let the interference produce the peak in probability that we require. We know that the amplitude may contain the sum of many waves. See what becomes of it when we add waves of different wavelength that are all in step at the central point, the point close to which we expect to find the particle."

Below the first wavy line appeared another, very similar except that the peaks of this one could be seen to be a little closer together. This second line moved upward to merge with the first, and in the center, where their peaks coincided, the total was greater than for either alone. On either side the two waves gradually got out of step until, some way off to the side, a peak of one coincided with a trough of the other and the waves canceled completely.

"You can see that adding in a second wave has increased the amplitude and hence the probability in the center and reduced it farther out. We cannot get the effect we want with just two waves, but there is no need to stop at two. We may allow ourselves as many as we need."

The different wavelengths of the two waves had caused them gradually to get out of step with one another so that eventually they canceled, but it did not stop there. As Scrooge's eye scanned farther from the central peak he saw the waves get even further out of step, so that eventually they differed by one whole wavelength and the peak of one coincided with a *different* peak of the other. Once again they added to produce a higher probability. Yet farther out there was another cancellation, then another maximum, and so on, the sequence repeating again and again along the extent of the wave.

To counter this repetition of the central peak more wavy lines appeared, each one rising up toward the original and adding to it. All were centered at the same position, so every one of them added at the center to increase the maximum in the probability at that position. However, because there was now a considerable spread of wavelengths, different waves got out of step with one another in different places. Scrooge could see that, away from the central region, there was nowhere that they did not cancel. Now the total amplitude was large near the middle of the mirror but fell away to a negligible value on either side, and there were no longer any regions of high probability away from the center.

"That is a wave packet. A bundle of waves that conspire together, through interference, to make detection of the particle probable only in a certain region. The location of the particle is no longer *totally* uncertain. There is only a limited range of positions within which it is likely to be found. This still gives *some* uncertainty in the expected position of the particle, but not a lot. The uncertainty is much reduced. Here is a question for you to consider. How did I make this a localized wave?"

✉ SUPERPOSITION OF WAVE FUNCTIONS ✉

All "particles" have associated waves. That is a basic and inescapable rule of quantum physics. It is the way the world works. The wave gives the probability distribution, which tells where you are likely to detect the particle. If the position of the particle is to be well defined, then this probability distribution will only be large over a small region, so that it is improbable that the particle may be observed anywhere else.

A wave with a single, definite wavelength is not at all localized. It just goes on and on unchanged, as illustrated by either of the two waves at the bottom of the diagram. When you add two waves with different wavelengths, then you can arrange that in some

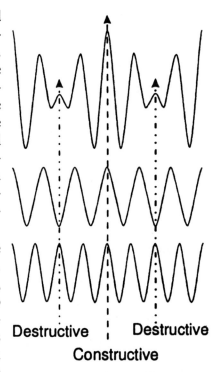

Destructive   Destructive

Constructive

places they interfere *constructively*; that is, they add up to give a larger amplitude there. As the wavelengths, are different they will gradually get out of phase, so that in other places they interfere *destructively*; that is, they cancel one another.

The amplitude of the total wave now varies, and the probability density is greater in some places than others. With just two waves the pattern will repeats and the waves come back into step again; but with a whole range of wavelengths it can be arranged so there is only one region of constructive interference and everywhere else the amplitudes are canceled out by destructive interference.

"I think you did it by adding together a lot of waves for some reason," answered Scrooge, who still felt a little vague about this process.

"Indeed I did," replied the Clown heartily. "I added together lots and lots of waves with different wavelengths. You saw me doing it! This way you can arrange that the wave function is localized in a central region where every wave has a maximum at the same place as every other. Because the wavelengths of these waves are all different they will gradually get out of step as you move farther away from the central region; and with enough waves, you can arrange that the different waves cancel one another out at any point away from the center. Since the wavelength of any wave gives the momentum of the particle, this localized wave function includes a whole range of momenta which the particle might have. Adding in these waves has made the predicted position of the particle more definite, but as a consequence its momentum is less definite than it was. There is no way round this. It always happens. If the position becomes less uncertain, the momentum becomes correspondingly more uncertain. That is certain. It is not possible to measure momentum and position exactly at the same time. You will always have some uncertainty in both position and momentum, and the product of the two uncertainties can never be less than a certain value fixed by Nature, a Universal constant which you call Planck's constant. There is a trade-off between size and momentum; and if you make a more careful measurement of one quantity, then the other becomes that much more vague and fuzzy."

The Spirit swung round and held up a finger in admonition. "You must not allow yourself to be confused by the word 'uncertainty' and to imagine that a particle does have a definite position and momentum but that you just do not happen to know what they are. That is not the way of it at all. If you try to force a particle into a tight position it gets momentum. In a tight position some people get religion, but a particle gets momentum. By your everyday standards the effect is very small, as Planck's constant is quite tiny. To appreciate what happens you must see it from close up, *very* close up. I ask for a volunteer from the audience."

Scrooge looked around him. As far as he could see, apart from himself the theater was deserted. Amid a burst of distinctly artificial sounding applause he heard the Spirit call out. "Scrooge, *come on down!*"

Impelled by the clapping around him, Scrooge left his seat and began to walk down the aisle toward the stage. At the same time the Clown leapt off the stage and began walking up toward him. Scrooge was struck by the thought that his mentor's vault down from the stage looked a deal farther than his previous leap onto it. Turning his head to look around him he saw that the rows of plush seats now seemed strangely large upon either side, as they might appear to a young child. With every step he took, the rows of seats

⫷ UNCERTAINTY RELATION ⫸

The quantum amplitude describes the best information you have about any particle and gives the only predictions you can make about future observations. The Uncertainty Relation says that a particle that which has a small uncertainty in its position must have a large uncertainty in its momentum. Here the word *uncertainty* is used to describe the extent of the region for which the probability of finding the particle is high. If this region is small, then the position of the particle is well defined and the uncertainty in its position is small. We have seen that a narrow probability distribution can only be obtained only by adding many waves of different wavelengths, and as given by the de Broglie Relation this means different momenta. There is thus an uncertainty in the momentum. The narrower the local peak in the probability distribution the greater is the range of wavelengths (and momenta) needed to make the waves get out of phase more rapidly. This gives the final relation.

$$\Delta x \Delta p \geq \frac{\hbar}{2}$$

where Δx is the uncertainty in position and Δp the uncertainty in momentum. $\hbar$ is a modified form of Planck's constant.

---

became larger until he was looking up at them from below. He found that walking was becoming difficult and turned his gaze toward his feet. As he had shrunk, so his feet had sunk farther and farther into the pile of the red carpet, and now it was up to his ankles, his feet becoming tangled in the twists. He stopped walking and stood in bemused resignation as the red tufts rose higher about him, now waist high, now towering over his head.

From this point his descent was swift. The carpet tufts became huge twisting cables, every filament a great curving cylinder. Surface irregularities and defects of the slick artificial fibers became evident, became all that filled his vision. They expanded into a glimpse of long arrays of fuzzy atoms such as he had been shown by the Shadow of Entropy, and then they too were gone. He was in a place that was no place. In a diffuse fuzzy grayness, devoid of any recognizable feature, save one. Approaching him through this fog of uncertainty came his companion the Clown, smiling quite as brightly as ever. He stooped to pick up something from his surroundings and abruptly threw it toward Scrooge, crying "Here, catch this!" Scrooge flung out a hand and felt it close over a small object. With it clasped inside his hand he could not tell what he had caught, but it felt to be alive for he could feel it fluttering within his clenched fist like a trapped bird.

"Now squeeze it, squeeze it as tightly as you can," commanded the Spirit, somewhat callously it seemed. Scrooge obeyed and found that as he closed his hand more tightly, so the hidden object fluttered and raced to and fro more violently than before. The outward pressure it exerted on his hand was so great that he had not the strength to press down upon it any further. He opened his hand to see what manner of thing it might be, but saw only a haziness which departed at speed.

"What was it then?" he asked.

"An electron. That is how electrons and other particles behave. When they are squeezed into a small space they are forced to have momentum and to rush about. The more tightly they are constrained the more active they become. As I said, they get momentum."

"But from where do they get this momentum?" asked Scrooge in some confusion. "I was told that momentum is conserved, and that if one thing gains more momentum this must be balanced by an equal momentum which is lost by something else."

"Well of course that is broadly true. Why certainly it is. I would not wish to deny one of the great conservation laws. Where would we be without them? You can be quite as certain of the conservation of momentum as you can be of anything, but *no more certain*. Even such Universal Laws must live with Uncertainty. I don't deny that momentum is conserved, by and large. Particularly large. When you look at the broad picture and do not take a restricted local view, then objects and particles are not tied down in position, and so the uncertainty in their momentum may be very small. In such circumstances momentum is indeed conserved, just as you have been told. If, however, you limit your vision and require that a particle be confined, then its momentum will fluctuate.

"You have seen photographs in newspapers, have you not?" he asked abruptly. "They look something like this." He made a sudden lunge among the indecipherable shapes that littered the scene and brought out a crumpled newspaper. This he smoothed out to show a picture of the Clown himself. On close examination it was quite indistinct, being made up from a number of black dots of different sizes which formed the light and shade of the picture. "Nature draws her picture of reality in a like manner, but Nature does not produce still-life paintings. Her pictures show not only the position of Being but also the motion of Becoming. Her canvas shows position and momentum equally. In short, the space in Nature's painting is *phase space*, where the width across the canvas portrays distance, but the height represents motion. On such a canvas the pictures are also made from dots."

From out of the vague haze that surrounded them came floating a large mahogany desk and behind it, apparently quite unsupported, a large chart. The Clown sat behind the desk, in a high-backed swivel chair, like any dynamic executive—at any rate, like any dynamic executive with green hair, a huge red nose, and wide brightly colored suspenders. On the desk in front of him was a small sign that read "The chaos stops here."

Scrooge looked at the chart behind the desk and saw the distinctive ultrareality of phase space, as he had seen it before on Father Time's tablet. On this occasion the phase space picture, like the newspaper photograph, was made from dots. They were not featureless points but little blobs which were all of the same size and strangely diffuse in appearance. They were all the same size in that they all had the same area, but they were of many different shapes. Some were approximately round, some were very long, but correspondingly narrow.

"You see the blobs, each representing the state of some particle which is included in the scene. Some are tall and narrow; these are

particles whose position is well defined and their momentum correspondingly vague. Other are very wide and low, these being particles whose position is really uncertain and their momentum well defined. Other spots are almost circular; these show particles whose position and momentum are equally well determined. *Equally well* also means *equally poorly*, since this is all the precision that is allowed. Each blob has the same area, however it might be proportioned. The spot may mean an accurate position and a wildly fluctuating momentum, or it might be a very rough position and a momentum that is well known. In all cases the area, the product of the two, is given by that Universal constant, which you call Planck's.

"These little spots tell you the combined uncertainty for the position and the momentum in any direction. Nature is three-dimensional—three-dimensional at the very least—and this fuzziness holds independently for each of the three dimensions of space. Along one direction you have an area of confusion whose size is given by Planck. If the object you consider can move freely on two dimensions, there is a Planck's worth of uncertainty in each direction. Its overall uncertainty is as thick as two Plancks.

"You have been shown before that Nature does not operate in space alone but in the combined space of *space-time*. Along this extra axis, the axis of time, Nature shows a corresponding degree of uncertainty. As with position and momentum, so it is with time and energy. Any particle shows a combined fuzziness in time and in energy, which is given by that same Planck's constant. This means that a particle which has a long and peaceful life, that is in a state that does not change at all over a very long period of time, will have an energy that is well defined. For such a particle the Law of Conservation of Energy is satisfactorily obeyed, and the sway of the Mistress is fully recognized. If the particle is short-lived, however, and if it is a fly-by-night affair that is here one moment and gone the next, then the conservation of energy is not for it. Its energy may show great fluctuations, and it may do things that energy conservation should by rights prohibit.

"Particles are the pickpockets of the universe. They are forever stealing energy from Nature. They cannot actually keep it for long before they have to give it back again, but fluctuations in a particle's energy are always present. The shorter the time, the larger the fluctuation; that is the rule. This continual thievery of energy allows a particle to be contemptuous of many constraints. To it the forbidden is now allowed. It may boldly go where no particle should be able to go. Look at this potential barrier."

⚜ Energy Fluctuations ⚘

As well as the uncertainty relation between momentum and the position of a particle, there is an equivalent relation between particle energy and time.

$$\Delta E \Delta t \geq \frac{\hbar}{2}$$

where Δt is the uncertainty in time and ΔE the fluctuation in energy. The consequence of this relation is that for a sufficiently short time (Δt) the energy of a particle may fluctuate by an amount of energy (ΔE). Though energy is still conserved on average in the long term, in the short term it is *not*. The amount of energy available fluctuates for short periods, and the shorter the period the greater can be the fluctuation.

"I beg your pardon!" exclaimed Scrooge. "What may that be, if I might ask?"

"Of course you may ask. You may always ask. I might not always answer you, but this time I shall. A potential barrier is a wall for particles. You may speak of a solid wall, but what is a solid? It is a lot of atoms close together, and other particles may or may not be able to move among them. It is energy which controls how and where a particle may go and potential energy that defines the terrain that it must cross. A potential barrier is a region where the potential energy of a particle rises abruptly as it enters the barrier. If a particle should enter the barrier region, its potential energy would become greater than its total energy; and so you would say that the particle cannot enter. The total energy of any object is the sum of its kinetic and potential energies, so if the potential is greater than the total this must mean that the kinetic energy is negative. You know that the kinetic energy is given by the square of the particle's velocity, and squaring any number gives a result that is always positive, whether the number itself be positive or negative. Obviously kinetic energy cannot be negative, and so the particle may not enter the barrier region. That is why it is called a *potential barrier*," he finished rather breathlessly.

"Now look at the wave functions in the regions of different energy and see what actually happens!" commanded the Clown.

"How can I?" protested Scrooge. "I can see neither wave functions nor energy. I can see only positions, the positions which the particles occupy in normal space."

"Now don't be tiresome. You must know that you cannot *see* the positions of particles either. They are far too small. Indeed, you never actually *see* the position of any object. What you see are photons, the particles of light that have scattered from the object and then enter your eye. You have learned to interpret this as *seeing positions*, but that is an *interpretation*. Babies cannot see the positions of things; they have to learn the skill or, rather, the convention. Here you will just have to unlearn that skill, for it is not appropriate. You know very well that you are not here in your normal body. You have been told often enough that this is a vision, so try to use a little. Vision, that is. Here it is more appropriate to see energies and wave functions, so do use your imagination. Even better, use mine!"

Scrooge felt as if his eyes had suddenly been opened afresh and saw as if for the first time. His vision blurred and then cleared, though on clearing it revealed a strange new way of seeing. He saw a hollow with walls on either side, but somehow knew that what his eyes interpreted as high or low was in fact the peaks and troughs in the potential energy. Where he saw a hollow, there the potential energy was low and particles would have the greater kinetic energy. Where he saw a wall he knew that he looked upon a region where the potential energy was high and particles would be rejected. Over the entire scene he saw a clear level surface, like the surface of a crystal pool above a craggy bottom. This surface he knew to mark the total energy of a particle. It was always at the same unvarying level because energy was constant and conserved. The depth of the potential level below the flat surface gave the full measure of a particle's kinetic energy. In some places the bottom of this energy pond was far below the surface, and the kinetic energy high. As the bottom rose closer to the surface, so the kinetic energy became the much less. Around the edge of the pond, the bottom was a bottom no longer but rose around the pond to contain it, forming narrow walls like the walls of a child's paddling pool.

Within the pool Scrooge could see waves. Waves on a pond might seem normal enough, but these were not on the surface, were not apparently in any way related to the surface but, instead, spread independently throughout the pond in ripples of fluctuating amplitude, which Scrooge now found himself able to see. The wavelength of these probability waves varied from place to place. Where the pond was deep the wave's peaks were tight together, showing that the wavelength was small because the momentum and kinetic energy were high at this point. Where the pond was shallow the waves were that much more widely spread and the wavelength longer. Where the bottom rose up to become a wall, then it rose above the surface and so obviously there would be no wave at all. *But there was!* Within the wall Scrooge could "see" the wave continue. It no longer rose up and down in peaks and troughs. It had no wavelength as such, but there was a wave nonetheless, an amplitude that faded steadily away as it soaked into the wall.

The wall was thin, and the decaying wave had not vanished completely by the time it reached the outer edge of the wall. There was still some amplitude left, tiny but still visibly present. Outside the wall the potential energy was again low, was again well below the total energy level for the particle. Here Scrooge could see the amplitude fluctuating again. It was now so small as to be scarcely perceptible but had again the short wavelength that denoted a significant amount of kinetic energy.

"Watch what happens to electrons in this container. It makes a good box, does it not? The walls are so high the electrons do not have enough energy to cross over them." The Spirit reached into his pockets, drew out what looked to be a handful of dust, and threw it in the pond. The waves became more intense, and every time Scrooge looked closely he could observe a selection of electrons scattered here and there within the enclosure. They were now in one place, now in another, as he made a series of observations; but always they were safely contained within the box. Until, that is, the moment came when he looked and saw that one electron had escaped. It was outside and departing rapidly.

"There you see it. Quantum paddling pools leak! The walls may be too high for the quantum wave to flow over the top, but it can *tunnel through* the wall. The amplitude falls quickly as it penetrates, but if the wall is thin then there will be some amplitude remaining on the far side. The amplitude outside will be small, and the probability of finding a particle outside will be small. It is small, but it is not zero. It is *possible* for a particle to escape from a sealed room and so of course they *do*. It is such barrier penetration that allows many radioactive decays to happen.

"As is the case with those tunneling particles, many particles will steal a little energy or momentum from the universe. It is a common thing. Usually they do not take very much, and they always put it back. Fluctuations are allowed, but the final totals must balance. In the longer term, energy and momentum are conserved.

"Some particles go further. They do not just steal a little energy. They steal all they can and all that they have. They steal the very most that they possibly can. They steal *themselves!* They steal *existence.* Such is the behavior of *virtual particles.* You have learned already that particles have a rest mass. Mass is energy, so this rest mass is the minimum energy which they must have in order to exist. It is their entrance fee into the select society of reality. A *real* particle is a fully paid up member: It has free and absolute possession of the energy it needs. If it is moving it has a greater energy, but its rest mass represents the lowest it may have, the energy of a particle that is not moving at all. The entry conditions for existence are clear enough, save that the particles themselves are not so clear. Are they moving or at rest? Uncertainty does not allow you to say for sure. Do they have energy or not? Fluctuations make this unclear also. Large fluctuations may happen only for a short time, but for some very short time a particle may have a fluctuation in its energy which is as large as its rest mass. Even though it was not there at all it may suddenly exist for a short time, though its time will be very short. From nothing it comes, to nothing it must once again return and that right smartly."

"I find this almost more difficult to accept and comprehend than anything I have been told before," said Scrooge, "but even if it be so, does it matter? Such a brief existence as you describe can surely have no significance, no impact upon the real world, that solid world which I normally inhabit."

"There you are mistaken. Truly, wonderfully, totally mistaken. These virtual particles, these fleeting scraps of transient borrowed reality, have the greatest impact upon the solid world that you could imagine. It is virtual particles that support and bind your real world together. Without them there would be no solidity at all, only particles moving past one another without any interaction. Tiny, independent rips in the untroubled fabric of reality, they would be rips that pass in the night with no awareness of one another. The world is solid and substantial because particles are bound to one another. There are interactions between atoms, there are forces. Within a solid the particles live in a force field. And what is there in this force field apart from particles?"

Scrooge had a vision of a field, a pretty grass-green meadow with a neat fence and white painted gate. Scattered around the field were a number of improbably fluffy white sheep, each grazing in his own stretch of pasture.

"Wrong sort of field!" exclaimed the Clown, "we want potentials, not pasture; and our flock is to be a flock of atoms." The fence and gate vanished, the green grass faded to neutral anonymity, and the fluffy sheep transformed to fuzzy atoms, held in a regular array by the forces which they exerted on one another. Between the atoms the field appeared featureless; but on closer examination it was not totally empty. There was something there. Something filled the space between the atoms, something that *was* the force field itself, surrounding and containing the atoms and holding them together. Scrooge's view continued to change, and he saw waves such as he had seen earlier in the potential pool he had been shown. Now there was no contoured bottom, no underlying potential, but only a uniform background on which were superimposed such decaying waves as had tunneled previously through the potential walls. There was a confused mixture of waves, each and every one centered on a particle and extending beyond it, becoming fainter and weaker as it went.

"You see the amplitudes of tunneling particles everywhere that you look. In this case the tunneling particles are photons, created by the electrical charges of the electrons and other charged particles within the field. Fluctuations in the energy allow photons to live for a brief time beyond their nonexistent means, but there must be electric charge to act as progenitor and midwife to their transient being. The amplitudes for these photons are much as you saw for particles tunneling through a potential wall, a place where they had no right to be because they did not possess enough energy. These photons may be out in the open, with no potential barriers in view, but they still have no right to be there because they carry energy which they do not have. They are living on borrowed energy, which is the same as borrowed time, and they cannot roam very far from the electron or atomic nucleus that is their origin. The probability of encountering such a particle falls rapidly as the distance from its source increases, but within that distance it may encounter another electric charge, carried perhaps by another electron. It may cast itself upon this charge and end its brief existence. In so doing the photon has been exchanged between the two electrons. It began its life with the electric charge of one and ended it upon the electric charge of another. In the process it has carried its stolen energy and momentum from one to the other, and so has produced an interaction.

"As is always the case, this photon exchange contributes an amplitude, a part of Nature's vision of the system. It is in the nature of amplitudes to be positive or negative, to add or subtract; and the overall effect may be to attract the two electric charges to one another or to repel them. Which it is to be depends on the electric charges involved. Similar charges repel one another, opposite charges attract; but in either case it happens through the exchange of photons between the charges. I said that in this case there were no potentials present. I lied. The photons *are* the interaction, and so they provide the potential. There is no other.

"You asked, on one occasion, how we could say that all different forms of energy were the same."

"Yes, I remember," agreed Scrooge. "I was talking to the Mistress, who had told me that kinetic energy, gravitational potential energy, electrical energy, and heat were all somehow the same, despite evidently being quite different."

"Well now you can see that they are 'somehow the same' for the sufficient reason that they *are* the same. In each and every case you are talking of the same thing, the energy of particles. A moving particle has more energy than one at rest, though that has its rest mass energy, which is usually large. A virtual particle also has energy and momentum, though in proportions different from those in a 'real' particle. The kinetic energy of a large body is just the energy of all the particles within it as they share the common motion. Heat is the energy of particles as they move, each in a different direction within a large body. Electrical energy is the energy of the photons that make up the electric field, the method by which electrical interactions are transmitted from one charge to another. Gravitational potential is similar to electrical potential, but with different particles exchanged, in this case something called a graviton. All of those forms, which seemed so different, may every one be seen to be the energy of particles. All energy is the same. It is just—energy.

"Everything which you see and work with in the world about you is composed of particles. Pretty well everything that affects you is made from but two different types, electrons and photons. It is the interactions between these two which make up the material world. If you ask me, 'What is matter?' I must answer that it is electrons and the positive charges in the nuclei of atoms and the photons exchanged between them—that is matter. Photons are the particles that make a beam of light; they are also the particles that hold together a rock or an iron bar. How would you reply if I were to tell you that an iron bar is held together by chains made of the stuff of moonbeams?"

"You would do better to say moonshine!" retorted Scrooge. "As that is the most obvious nonsense."

"As I have remarked before, you must try not to be obvious. The remark is in fact nothing but literal truth. A moonbeam is a stream of photons, of real photons. Iron is held together by electrical forces, by virtual photons that exchange energy. A photon is a photon. Some may have different energies from others, but they are the same thing. Electrical forces tie particles together to build the structure of the world, and the building block which they create is the ATOM."

CHAPTER 11

No Atom of Doubt

"The world is made of atoms," stated the Clown, adopting an authoritative pose. "Of that there can be no atom of doubt. Every material thing is composed of atoms and to produce most of the matter in the world only a few types of atom are needed, but they combine together in such wondrous variety as to make all the myriad of substances which you see around you.

"All atoms are built upon the same design. We begin at the very beginning. We begin with a nucleus." The Spirit reached within his all-providing pockets and brought out something too small for Scrooge to see. Small though it was, it seemed to be very heavy, and he lifted it with difficulty. "Here we have a nucleus. There is a nucleus such as this in the center of each and every atom. It is very small, it is very heavy, and it carries a positive electric charge. For most purposes that is all that is required of it, since this is enough to ensure that electrons, which carry negative charges, are attracted to it. When I say the nucleus is very small, I mean *very* small, some one hundred thousand times smaller in diameter than the atom itself. The nucleus is but a tiny speck buried in the very center of an atom, but despite this it is so heavy that less than one thousandth of the mass of the entire atom lies outside it."

"Surely that must be wrong!" protested Scrooge. "How could something so small be so heavy? It is large things that are heavy, surely."

"No, it is quite correct," replied the Clown, "and please stop calling me Shirley! In the case of particles, *heavy* and *small* do go together, see for yourself!" The Clown set down the tiny thing that he had been holding and became momentarily blurred as he turned an abrupt somersault. When he was again upon his feet Scrooge could see that he was now dressed in the leopard skin of a circus strong man, though the effect was spoiled by the fact that it was fully as baggy and loose as his trousers usually were, while thin arms like pipe stems stuck out on either side.

He bowed to his audience, even though this consisted of Scrooge alone, and walked across to a rack which had not been evident before. On this were ranked a series of somewhat indefinite-looking balls, ranging in size from a large one that bulked one half the height of the Clown himself, right down to a tiny one like a large pea. He adopted a series of extravagant poses, flexing his arms so that implausible biceps popped out at intervals. He confidently reached for the largest ball of all, and bending his knees for a more secure lift he picked it up. He did this with little apparent effort and soon transferred the great ball to one hand, then balanced it on the tip of one outstretched finger, casually passing it from hand to hand before he set it back. Smiling confidently he moved to the other end of the line of balls and casually picked up the tiny one. Abruptly he pivoted ludicrously about one foot, while the other foot soared involuntarily into the air as his hand crashed down to the floor. He struggled and cavorted, flinging himself around, but anchored at that one spot where his hand was pinned to the ground by the immovable mass of the tiny globe resting upon it.

"Very dramatic!" acknowledged Scrooge. "But this goes no further toward explaining why a small object *should* be so heavy."

"You have it the wrong way round," responded the Spirit, quite calmly for someone in such an awkward position. "It is not so much that a small particle *need* be massive as that a massive particle is more *likely* to be small. To start with, it all depends on what you mean by small."

"Why, it is obvious what 'small' means," retorted Scrooge. "It means 'not big.' It means that the object, whatever it may be, does not occupy very much space."

"If, as you say," continued the Spirit, " a small particle is one that definitely occupies a small volume, then the uncertainty in its size is small and so its momentum must be large. That is how things are with particles in the quantum world. A heavy particle can more readily have such high momentum because momentum is proportional to mass. Heavy and small do go naturally together in this interpretation of small. You tend to think that heavy objects should be large because you are used to thinking of *compound* objects, things that are made up from lots of identical particles. Large objects will contain *more* of these particles than small objects and of course if there are more of them then they weigh more. For a *single* particle, however, small size usually means great mass. An electron is very much lighter than the nucleus of an atom and the electrons fill the *whole volume* of an atom, which is very much larger than the tiny nucleus. That is one meaning of size, and on that interpretation electrons tend to be much larger than nuclei. A different, and more conventional meaning is the intrinsic size of the particle. This is a definite property of the particle itself, rather than one forced upon it by uncertainty. In this sense the electron is a point particle and as such has *no size at all*."

"But," protested Scrooge. He paused for a second, replacing his automatic first choice of word. "Of a surety," he continued, "that is the exact opposite of what you have just been saying, is it not?"

"Perhaps it might seem to make nonsense of what I have just told you, but often the truth does seem to be nonsense. Truth is like that. It *is* true that the electron is point-like, in the sense that it has no known structure, no inherent size and form of its own. However, as I told you, an electron in an atom is *not* tightly localized. You know that it is somewhere within the atom and that is as precise as you can tell. In this situation the position of the electron is so fuzzy you cannot really say how large it is. In a sense, its fuzziness is its size, but not its true, internal *personal* size. The nucleus is much more localized, and it may readily be so because of the large mass of the nucleus. An electron has in its nature to be localized even more exactly than the particles that make up the nucleus. It *could* be so localized but only if it is given energy, vast amounts of energy, far more energy than is contained in the mass of a nucleus, let alone in the relatively puny mass of an electron. If an electron is accelerated to a speed close to that of light, then its energy will rise and rise and it can rise to many, many times the energy tied up in the electron's own rest mass. Its associated wavelength will then become very small, and then indeed an electron may be seen to be truly tiny, with no inherent size of its own."

The Clown, who had unobtrusively returned to his normal appearance, held up one hand with the thumb and forefinger pinched close together to indicate the tiny nature of the electron. He then spread his arms wide to indicate a greater size. "Protons and the neutrons, the particles that make up the nucleus, do have a finite size of their own, one which is not due solely to their fuzziness. However fast they travel, however their energy rises and their wavelength decreases, they are still seen to have this size. It all depends on what you mean by size. The nucleus has an inherent size much greater than the electron, but the quantum fuzziness in its position is sufficiently small to allow its true size to be seen. 'Size' can mean two rather different things, you see. Any particle has a quantum fuzziness and spreads itself over some region. This depends on the energy of the particle and will decrease as the particle gains more momentum. An object may also have an intrinsic size, an individual property that does not decrease as the object moves more quickly, apart of course from normal relativistic contraction. Such intrinsic size will usually arise because the object is built up from many particles packed together.

"Objects you are familiar with in your everyday life have size, but their size comes about because they contain many atoms. Each atom is in a different place within the object, and so together they

⊲⊙ Particle Size ⊙⊳

The size of a particle is not as simple a concept as you might think. Because of the inherent *spread* or *fuzziness* in a particle's position, as expressed by the Heisenberg Uncertainty Relation, a particle will effectively occupy a finite volume. In a sense you could say that is the size of the particle, as it is the region over which the particle "extends." In an atom the electrons extend over the whole volume of the atom, while the nucleus is localized to a relatively tiny region. The diffuseness of the particle depends on its momentum and decreases as the momentum increases. Heavy particles will have greater momentum for a given energy and so they tend to be more localized. In this sense *small* and *heavy* go together. If an electron is given enough momentum it can become very localized, effectively pointlike. This is not so for the nucleus, which appears just as large when it has a high momentum as for low momentum. The nucleus has a size that is intrinsic, a property of the nucleus itself, while the electron does not appear to have any intrinsic size at all. The higher its momentum, the smaller it seems to be. In general, those particles that do seem to have a definite size are compound, made up from other particles.

---

fill a considerable volume. Each atom has its own quantum amplitude, and for each atom this fills some small volume with a high probability for the atom's electrons being found there. For each atom it is a different small volume, and if there are many atoms, as there are in anything large enough for you to see, then these volumes all pack together side by side, and the object has size. The electron may in itself be tiny, it may perhaps have no size at all, but an electron has so little energy in an atom that the fuzziness in its position fills the *entire volume* of the atom. When the probability amplitude for a particle is large, then you could say that the region is filled by the particle. This is really all that you can ever mean when you say that a region is *filled by particles*. It is not true to say that atoms contain mostly empty space. They are completely full of electrons.

"It is now time we returned to where we left the nucleus." The Clown bent down and, with great show of effort, picked something quite invisible from the ground before him and, straining mightily, lifted it to shoulder height and appeared to set it on an invisible shelf at eye level in front of him. "This nucleus is a compound object. It is made from protons and neutrons. The proton and neutron in their turn are also compound. They contain other particles, the quarks. We may have more to say of them later; but like the electrons in the atom, the amplitudes of the quarks fill the volume of the proton or neutron and so give them size. When many protons or neutrons are packed together in a nucleus, then the nucleus also must have a distinct size, in the same way that large objects containing many atoms have size. As I said some time ago, all that we require of the nucleus is that it be small and heavy to make an anchor for the atom and that it have a positive charge to hold electrons to it.

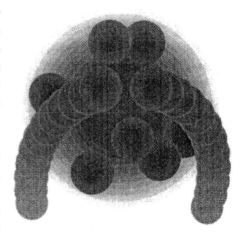

"Come! Watch me as I juggle in some electrons to form an atom," he continued, reaching into his pocket once again. He threw a handful of electrons into the air, or whatever it was that surrounded them in this strange environment, and began to juggle them in a tight, spinning circle around the place where he had set the nucleus. They became a blur above his head. Indeed they had never been anything but a blur. Scrooge could see the entire group as a fuzzy shape, no

more detailed than the hazy spheres that he had been shown by the Shadow of Entropy in an earlier vision. Was he never to see the structure of an atom in more detail? Was there no form or structure at all? He voiced these thoughts to the Spirit.

"As I have told you already, you must look at things in the right way. In space the position of the electrons is indeed fuzzy, and you cannot hope to see an atom as other than a hazy sphere, but there are other ways of looking. Look at *energy*, look at *amplitudes*, and then you will see more clearly." The Spirit hurled the fuzzy ball of electrons upon the ground, and at the same moment Scrooge felt once again the shift and clearing of his vision that he had experienced when he looked before at the quantum pool. He saw a depression in the ground before him: the flat ground that marked the constant plane of zero energy. The depression was circular, and its center was very deep. The edges curved down in a smooth trumpet shape toward a center too far down to see. Scattered round about he could see other trumpet-shaped depressions, identical to the first.

"There you see the potential wells produced by the electric charges of the nuclei at the center of atoms. An electron that approaches the nucleus will be attracted and its potential energy will fall ever more quickly the closer it gets to the nucleus. This fall in potential is called a *well* because it goes down so deep, and it is within this well that the amplitudes for the electrons must fit.

"The electron may be anywhere within the atom, and all possibilities combine together in the total amplitude which describes the electron. The various possibilities will interfere, and it turns out that only a distinct set of possible *states* remain. These are the possibilities that do not destroy themselves by interference. Only this set of possible states gives nonzero probability distributions for the electrons. All other possibilities are *not* possibilities because they have no probability: since Anything that has a probability of zero is not a possibility. To put it another way, they just can't happen. The situation is similar to the limited set of notes you may play upon any one pipe of an organ. The molecules of air within the organ pipe may move in all sorts of directions, but overall the waves in the pipe are restricted to give those notes to which the pipe is tuned. For electrons in an atom, each electron must be in one of these states, and the state fixes the energy of the electron it describes."

"The levels that you see show the energies that an electron within the atom *may* have, that it is *allowed* to have by the states available to it. Any electron in an atom will be in one or other of these levels. Usually it will be in one level and not in a superposition of them all because the levels are different in ways that could be detected. You only get a superposition of states when the states are quite indistinguishable from one another, or *degenerate* this is called. These levels within the atom give different energies for an electron, and it is possible to tell them apart.

"Atoms survive indefinitely, and within an atom the electrons stay for a long time in whatever states they may occupy. The states are consequently called *steady states*—states that are unchanging and so give little or no time information. The electrons within them are not restricted in time, and so the energy of each state is well defined.

"Any of the levels might be occupied by an electron. Each permitted state could hold an electron, and in different states the electrons may have different energies. Normally an electron will fall into whichever available state has the lowest energy."

In the depths of the potential well Scrooge could see the lowest energy level had been intensified, to indicate that it was occupied by an electron. He was aware also of a flickering over the surface of the surrounding plane, outside the atom's potential. The Clown told him that this was due to the photons present in the light that was shining upon the atoms. Abruptly he observed one of these photons to collide with the atom in front of him. The photon vanished, and he saw that a higher level within the atom was intensified now, showing that the electron had made a transition from the lowest level to this higher one. In the process the photon had vanished completely. It has been utterly absorbed and its energy used to raise the electron energy to the higher level.

"It is a consequence of the wave-like nature of the electrons. The range of probability waves is very limited in a way which must already be quite familiar to you for other waves."

The Clown reached into a pocket and produced a tiny whistle, on which he blew a thin high note. He then reached into the same pocket and drew out a long tube. A very long tube he drew it out and out until he had produced a horn, much like an old coaching horn, which was almost as long as he was himself. Putting this to his lips he produced a 'raspberry'-like sound. With a frown of concentration he drew a deep breath and blew a deep echoing note. He then blew a sequence of short notes, apparently as some sort of tune, but as every note was exactly the same it had a certain lack of variety.

He stood up, took a very deep breath and then, leaning forward, he succeeded in producing from the horn a higher ringing note. Flushed with this success he played a rousing hunting tune whose pace and delivery managed to mask the fact that the horn could still produce only two notes. At the end of this performance he bowed even lower and then addressed Scrooge once more.

"You see there how a different sort of wave, a *sound wave*, is limited by the form of its container. The small whistle gave a note of high frequency, a very high-pitched sound. The much longer horn

gave a correspondingly lower note, the frequency being given by the size and shape of the instrument. As you saw, or more probably heard, the horn is capable of giving more than one note but the choice is very limited and you may select notes only from the limited discrete set which is determined by the instrument. The molecules of air within the horn may move in all sorts of directions, but overall the waves in the tube are restricted to give those notes to which it is tuned.

"It is much the same for electrons in an atom, the atomic potential restricts and determines the frequencies of the amplitudes for each possible state, and the electron's energy is proportional to this frequency. Each electron must be in one or other of the states and the state fixes the energy of the electron it describes."

"This electron is now in an *excited state*, which simply means that its energy is greater than it need be. Such electrons prefer to sink back to the lowest state, the *ground* state, and in the process lose their surplus energy by emitting a photon. If the electron can do this, it is most *probable* that it will. Each of the electron states within the atom is but a single state, but any photon which is released could end up in any one of many, many different states, depending on the direction and motion of the photon. The emission of a photon is thus probable. It is strongly favored, *very* strongly favored. If the electron can decay back to the ground state then it probably will; and you have seen before that *probably* can mean *almost certainly*.

"I said the decay to the ground state is likely if the electron *can* decay. The electron started in an excited state and this is different from the ground state. The states are definitely, discernibly different, and so they are not part of the same superposition. The amplitude must really *change*, and this only happens if there is a transition, if something actually *causes* the state to change. When you saw an electron excited from the ground state by a photon, then it was the interaction with the photon which caused this transition. If the electron is in an excited state, then a similar interaction with a photon may cause it to fall back to the ground state, in the process emitting another photon to carry away the energy released. In the case of such *stimulated emission of light*, there are afterwards two photons of the same energy, the one that caused the transition and the one which was released to carry away energy from the electron.

"An electron in an excited state will be seen to fall back to a lower level even if there is no light shining on the atom to stimulate the transition. It might appear as if the states may change even when there are no photons present to provoke the transition, but in fact there are always photons. Even if there are no *real* photons present, no light shining on the atom, there will always *virtual* photons. The vacuum, everywhere, all of the time, is full of virtual photons, born of fluctuations in energy. These photons do not hang around for long, but they can be present long enough to cause a transition. What is called the spontaneous decay of excited atoms is in fact driven by such virtual photons.

"When an electron falls from its higher-energy excited state to one of lower energy, the transition is driven by the action of these virtual photons, which have no energy beyond what they may borrow as a quantum fluctuation. When the electron falls, it loses energy and this is carried away by a photon. The photon that is emitted has energy it may keep, and so it may be seen as a visible photon. As a consequence: Let there be light!"

From somewhere in the distant murk that surrounded them, a bright beam of light sprang up, shining toward them. The beam shone through a mist or cloud of fine droplets that gave a bright halo around the light. This was divided into bands of different colors, like a rainbow. The bottom half of this bow was not cut off by the earth, and so it formed a complete circle around the light. "There you see light, light of all colors. All the colors of the rainbow, in fact. The color you see depends upon the frequency of the light and upon the energy of the photons, since the two are related. If you should look at light that comes only from atoms of a given type, then the light is no longer 'all the colors of the rainbow.' Sodium atoms, which are often used in gas-filled electric street lights, give a very limited range of color."

The bright light in the distance changed from white to a distinct yellow, and the smoothly blending bands of color in the blazing halo which surrounded it all faded to black. All that remained were a few sharp rings with no apparent width. Most prominent were two bright circles of an intense yellow, two precise and closely spaced narrow hoops isolated by wide bands of darkness. "You see there a *line spectrum* for the light given off by an atom. The light cannot be just any color. Photons are given off only with whatever energy the electrons release when they transfer from one level to another. As the levels have distinct energies, the differences in their energies also give a distinct set of values. Only photons of those energies are produced, and consequently only the appropriate colors will be seen. Sodium light is striking in that almost all the photons come off with very similar energies. There are a few other colors produced, but they are very dim by comparison. By and large, the light is all yellow. This makes for an efficient lamp because most of the energy goes to produce photons which the eye may easily see. For relatively little cost in energy you get a light by which you may clearly see anything, any color you like as long as it is yellow. If there is only yellow light present, naturally yellow is all that you can see. Unfortunately, people became tired of seeing nothing but yellow. It made them feel blue, so eventually they saw red and started to use other types of light, such as mercury lamps. Their light comes from the decay of electrons in excited atoms of mercury."

The bright light changed again, now to a blue-white color. Again the surrounding halo was mostly blackness, but it contained bright, narrow circles of sundry different colors. "As you see, these lamps also give light of only a very few colors, but there is now a sufficient selection that they do give a rough impression of normal illumination."

The light flickered out, having served the Spirit's purpose. He pointed at the potential well, which Scrooge could still see before him. One of the many planes that represented the available energy levels within the atom was intensified to indicate that this was populated by an electron. As Scrooge looked on, this intensification, this highlighting of a particular level, flicked from one to another, back again to the lowest and then up to yet another higher level, illustrating the excitation and decay of the atomic states as the electron transferred from one to another, absorbing or emitting light in the process. The Clown gestured toward this sequence of transitions, which finally came to a halt with the electron indicated as being in the lowest level of all.

"I have been speaking of what happens to one electron in an atom. If there is only one it will fall into the lowest of all the energy levels within the atom. Most atoms, however, contain many electrons. The nucleus has a positive electric charge which is several times as large as the negative electric charge possessed by an electron. This means that the nucleus can attract that number of electrons before the atom is electrically neutral, at which point the electrons that it contains will screen the nuclear charge from attracting any more. Would you not expect that each of these many electrons should end up in the lowest level of all?"

"Well, I suppose so," answered Scrooge. "From what you have said so far it would seem only reasonable...."

"That was a rhetorical question," said the Clown severely, "I am looking for the *correct* answer, and that means I shall have to answer it myself. The answer is that the electrons will all go to the lowest level that they *can*, but they will not all be able to end up in the *very* lowest level because of the Pauli Exclusion Principle."

"And what, may I ask, is that?" said Scrooge.

"Why of course you may, of course you may! You did not have to ask though, because I was about to tell you anyway. The Exclusion Principle comes about because all electrons are identical. They are not just similar; they are completely identical. There is absolutely no way to tell them apart. As far as you can tell *they are all the same electron*. This makes them difficult to identify. Let me show you."

The flat plane of energy levels with its scattered potential wells faded from view, and Scrooge and the Spirit were again in a place of no discernible features. Through the colorless mist that surrounded them came floating a brightly striped fairground stall, which drifted to a stop beside them. On the front of the stall was a sign that read 'Find the Electron!' The Clown slipped inside and leaned confidently on the counter at the front.

"Let us see how good you are at observation. Try to predict the position of the correct electron." From under the counter the Phantom produced three cups and placed them upside down on the countertop. With a flourish he produced an electron, which he displayed to Scrooge. "Would you recognize this electron if you saw it again?" he asked.

"No, I would not!" The electron looked completely vague and featureless. "Can I mark it in some way?"

"Don't be silly, of course you can't. Now you see me put the electron under one of these cups and I put different electrons under the others. Now watch carefully!"

"Just a minute!" exclaimed Scrooge, who had been struck by a thought. "If, as you say, everything in the world is made from large numbers of atoms and the atoms are mostly built of electrons, then what are those cups made from? Will not the cups already contain many, many electrons, so that one, more or less, will be neither here nor there?"

"Why, you must surely realize that these cups are just *metaphors* for some form of potential well or state that can act as a *container* for an electron. The cups are but allegories, a little extra flourish added to make the demonstration that much more dramatic. See for yourself." The Clown turned over one of the cups and Scrooge could see, printed on the bottom, the words:

Made From Fine Bonus Allegory

The Spirit arrayed the cups in a line on the counter in front of him and began to switch them to and fro, his hand moving more rapidly than Scrooge could follow. Eventually the cups were again still, lined

up in front of him. "Right," said the Clown, "now can you tell me which cup 'your' electron is under?"

Scrooge had no idea whatever but indicated one of the cups at random. The Clown turned it over and looked inside. "Right," he said triumphantly. "You are absolutely correct, well done Sir! Perhaps that was a lucky fluke. Would you care to try again?"

There did not seem anything to lose, so Scrooge watched again as the cups were shuffled and chose one. His choice was again announced to be correct. He tried again and again, and each time he won. He began to watch much more closely that movement of the Spirit's hands and to choose cups which he was sure were not the one with which they had started, but each time he was announced the winner. He found the process to be increasingly frustrating. It is bad enough when you lose every time in such a game, but that would at least have the merit of agreeing with everyday experience. But always to *win*, that was unsettling!

"You cannot help but win, you know," smiled the Clown. "Whichever cup you choose it will always be your electron that is under it because, you see, they are *all* your electron. As I told you, they are not just similar, they are *identical*. They are all the same. Every electron is the same electron. When you have seen one, you have seen them all because they are all the same, they are all the *same electron*.

"Swapping the electrons between the cups makes no difference. There is no way to tell the different positions apart. I do not mean that *you* cannot do it, I mean that there is *no way* it can be done. All the permutations are the same, exactly the same, so whichever you choose it is *just the same* as any other. It is just as correct a choice, and if it is correct why should I not say so?

"If there is no way that *you* can make a choice, how can Nature do so? The answer, of course, is that She does not. The true state, the quantum state of the system, is not just any one of these permutations of the electrons, it is *all* of them! Nature holds a superposition of every possible combination and exchange of the electrons. She has to do this because of the general rule that anything that may be so *is so*. All possible states are present in the overall superposition. This is the principle that gives interference. It is the way that quantum nature works."

"But sh. . . . , it must be that when there many electrons are present there are a huge number of permutations all added in together."

"Why, of course there are. There are very many indeed. But what of that? It costs nothing. I have said that, because they are identical, swapping two electrons makes no difference. That is not quite true. There is one possibility, one very minor option. This is a rather subtle option that has no obvious effect, apart of course from the very existence of matter, of the world, and most of the Universe."

Scrooge tried to look unimpressed. "What then is this minor option?" he asked.

"Well, as you must now admit, electrons are identical. Swapping any two of them can produce no detectable effect whatsoever. This means that the *probability distribution* for the electrons must not change. There must be absolutely no alteration in the probability of finding electrons in different places or doing different things, as that would be an effect which you *could* detect.

"If you should swap two electrons *twice*, then obviously there can be no difference at all. You have put everything back exactly as it was before. Is there any possible change produced by swapping two electrons that would not affect the probability distribution and that would be undone when the electrons were swapped again? That is another rhetorical question!" he added, looking fiercely at Scrooge, who had not in fact made any attempt to speak.

"Yes, one option remains. There may be a change in the *arithmetic sign* of the quantum amplitude. It might change from plus to minus or from minus to plus. At any rate, it could change to the opposite. The probability distribution is given by the square of the amplitude, the amplitude multiplied by itself, and minus one times minus one is plus one. So of course is plus one times plus one. If the swapping is done twice and each time the sign changes, then in either case you are back to where you started, as is required.

"That is one option—that the amplitude changes sign. The other option is that it doesn't. Both options have been taken up by different sets of particles, and particles may be divided into two classes according to their choice in this matter. One class of particles are called *fermions*, and electrons fall within this class. They change the sign of their amplitudes when two of them are exchanged. The other class are called *bosons*. Photons fall within this class and do not change their amplitude in any way when two are exchanged. The particles of each type *always* show the behavior appropriate to their class whenever they are interchanged. Particles have a truly rigid class system."

"Very interesting I am sure," said Scrooge in a voice of no enthusiasm, "but what, if any, is the significance of all this? Does it really matter?"

"Does it matter, he asks," responded the Clown, casting his eyes up to what might well have been Heaven if they had not been in such an abstract and formless place. "Yes, it does matter. When two fermions, such as electrons, are swapped, then there is a change in the amplitude for the state that describes them. It may not be a striking change, but it is a change and it must *always* happen. This is no great problem if each and every electron is in a different state; but if two electrons are in the same state then swapping them really is *no* different at all. The state tells you all there is to know; and if two electrons are in the same state and you swap them, all you can say is that two electrons are in the same state. Where is the change? Where is the difference? There isn't one, but it is in the nature of electrons, part of their union rules if you like, that their amplitude *must* change sign. The only way that anything can be exactly the same and still change its sign is if it is *zero*, if it does not exist! The final conclusion of this rather subtle and logic-chopping argument is that two electrons *cannot be in the same state*. That is the bottom line. That is the *Pauli Exclusion Principle*, which says that two identical fermions may not be in the same quantum state."

"If electrons are so difficult, or even impossible, to tell apart, how can you be sure how many are in any state?" asked Scrooge.

"The fact that the particles are identical does not prevent you from knowing how many there are," replied the Clown tartly. "If you are dealing with identical twins you may not be able to say which is which, but you can certainly say whether one or both is in the room with you. 'Identical' humans are always somewhat different, whereas electrons are truly identical, but even so it is perfectly possible to say *how many* electrons you have in a state, though you cannot say which is in what state. The *number* of electrons is quite definite, and there is nothing vague and uncertain about the Pauli Principle. It does not say that electrons would rather not be in the same state, or that they would find it quite difficult to be in the same state. It is completely unequivocal."

He gestured to the gray mist around them. Before Scrooge's eyes a message appeared in letters of fire:

Two electrons may not be in the same state.

"That is the rule. It holds absolutely and invariably for electrons and other fermions. For bosons it does not hold. Photons do not mind being in the same state as one another, indeed they positively like it. Bosons are inherently gregarious and love to get a cosy group together, all huddled in the same quantum state. For fermions this is impossible."

"That does sound quite significant," admitted Scrooge, "but

would it really matter if electrons could gather together in the same states. Would I notice the difference?"

"I rather think you would. Let me show you a demonstration of a world without the Pauli Principle, and see if you can spot the difference." Abruptly they were transported from the gray, formless space that they had been inhabiting, and Scrooge found himself standing in a quiet country lane, where it ran alongside a small wooded copse at the corner of a field. Overhead the sun shone brightly in a blue sky, and the hedgerows were full of wildflowers and loud with the hum of insects moving busily among them.

"It all looks very normal to me," said Scrooge.

"Of course it does, it *is* normal. Nothing has changed yet!" said the Spirit impatiently. "See now what would happen if the Pauli Principle did not operate." There was an intense blaze of light around them, and Scrooge saw everything crumple and collapse. Every object visible sank into itself, collapsing to a gray, dusty uniformity, dark and amorphous. He felt himself sink into what remained of the previously solid ground, which now yielded to him like some featureless quicksand. The vision blended imperceptibly back to that place of fuzzy grayness in which the Spirit had earlier chosen to talk to Scrooge about quantum processes.

"The Pauli Principle lies at the heart of all the variety, all the diversity which you see around you in the material world. It is at the heart of Chemistry. It is even at the heart of the very solidity of the material world. Why do you think you cannot walk through solid walls?" he asked abruptly.

"Why, because they are solid," answered Scrooge, though he had a feeling this answer might not be acceptable.

"Yes, but that just means that you cannot walk through them. Why can the electrons in your body not mix and blend with the electrons in the wall, allowing you to slip easily within it? There are, after all, many electron amplitudes within each atom, so why should there not be a few more? The electrical forces do not provide an answer. They are largely neutralized because the positive charge on the nucleus is canceled by an appropriate number of negative electrons. But what force remains is, if anything, still attractive and would tend to pull your electrons into the solid. Do not bother to reply, by the way. I am just about to tell you the answer.

"It is not easy to compress a solid. Take a block of iron in your hand and squeeze it hard. You will have very little effect. This is despite the fact that the electrical forces within the solid are tending to help compress it further, but both their efforts and yours are resisted by the Pauli Exclusion Principle. Each atom in the solid contains electrons, and within each atom the electrons are in appropriate states. As the atoms are identical, the electrons in each are in the same states. This sounds already like a violation of the Pauli Principle—if electrons in different atoms are in the same state. The Pauli Principle is not just about electrons being excluded from the same state *within the same atom*. They are not allowed in the same state, without reservation. However, the same state in a different atom is not quite the same state, because the different atom is not in the same place. If it were, the states would be identical and the Pauli Principle would operate, so it has the effect of forbidding atoms from occupying the same space. In general, atoms will *not* pass freely through one another; indeed there is considerable resistance, and solids are not easy to compress at all, let alone down to the size of a single atom, which would be a possibility if they *could* interpenetrate freely. This is all because of the Pauli Principle. There is no other reason.

"This effect is quite distinct from the 'uncertainty' effect, which gives an individual atom its size. The electrons in an atom cannot be constrained into too small a size, because this would give them too much momentum and so they would have more kinetic energy than is conveniently available. This process would not prevent all the atoms in a solid from being in the *same* position because the electrons would be no more localized with the atoms in one position than if they were in another. The uncertainty relation, which connects small size and large momentum, gives the *size* of atoms; but it is the Pauli Principle for fermions that prevents the atoms from all being in the same place. The uncertainty relation would apply equally to bosons, but the Pauli Principle would not.

"Atoms are constructed from fermions, above all from electrons. From these atoms are constructed the almost infinite variety of matter which makes up the world. This variety is possible because of the Pauli Principle. It is because of *Chemistry*."

❦

CHAPTER 12

Per Amplitude Ad Astra

The Clown held up a brightly wrapped parcel, neatly bound in ribbon and finished with an elegant bow. He unfastened the bow and unwrapped the object to reveal the fuzzy electron cloud of a typical atom.

"The atom lies at the threshold of the quantum world. It marks the size at which quantum effects become obvious. Objects much larger than atoms may be

discussed in terms of classical physics. This has worked very well on an everyday scale, but on the scale of atoms or smaller it fails. There are two different paths we might follow from this point. We might go on down, descending ever farther into the realms of the quantum, going to smaller dimensions and higher energies."

The diffuse cloud of the atom which he was displaying seemed to explode suddenly outward, expanding to envelop both Scrooge and his companion. When the outer limits of the atom had passed completely out of his sight and they were deep within its very heart, Scrooge saw that the Clown was holding another tiny parcel. Its wrapping paper bore the repeated pattern

$$^{12}_{6}C$$

"Here, nestling in the heart of the atom, you have the nucleus, whose nature serves to define the type of atom that contains it. It is far removed from the typical scale of the atom, both in size and in energy. The nucleus is some hundred thousand times smaller in diameter than the atom itself and the energy of the particles within it some hundred thousand times greater than the atom gives to its electrons. That is without question a big difference in scale. This nucleus, like the atom that contains it, is a compound body."

The clown unwrapped this new package to reveal the portrayal of a nucleus, blurred in appearance as were all objects on this level of existence but visibly containing many other particles crowded together. As he brought these forward for Scrooge's inspection he saw them as a host of other and smaller parcels, whose wrappings bore either the symbol p or n. "The nucleus is composed of *nucleons*, either *protons* or *neutrons*. They also are compound, containing yet different particles."

The Spirit unwrapped one of these parcels also, to show that in its turn it contained three others, their wrappings decorated with the symbols u and d. "These are quarks. Each proton or neutron is a composite of three quarks."

"And I suppose that the quarks are in turn made up from other particles, and they in their turn from others, and so on," suggested Scrooge, who felt he was getting into the spirit of the thing.

"I cannot tell you," replied the Spirit calmly.

"Why ever not. Don't you know?" asked Scrooge.

"Perhaps, but your author doesn't. The quarks seem to be on the same fundamental level as the electrons themselves. There is no evidence that electrons are composed of anything else. Electrons, quarks, and photons all seem to be as fundamental as one gets. This is not to say that they cannot be seen to 'contain' other particles. The vacuum may contain virtual particles in combinations such as an electron with an anti-electron. You might expect that if a particle is truly fundamental, if it is not a compound state of other particles, then it should contain nothing. If it contains nothing, then it may contain the virtual particles that are contained in the vacuum, because the vacuum *is* nothing. This is no intrinsic property of the particle itself. In a sense you might say that photons contain pairs of electrons and anti-electrons. Because photons are the carriers and agents of the electrical field they provide something to which the electrically charged virtual particles may connect, but are these virtual particles *constituents* of the photon? Not really. In this sense you might say that every particle is composed of every particle, including itself!"

At this point the Clown swung round to direct an inquiring glance at Scrooge. "Do you really wish to pursue the nature of particles to this intertwining level of existence,[2] or shall we rather take the other path and move upward from the level of atoms to investigate the features of the larger world which you inhabit?"

"Yes," said Scrooge rather weakly, "I think we had better!" No sooner had he spoken than they again stood by the flat plane that represented the uniform background level of energy, with the potential well of an atom curving downward in front of them.

"The nature and behavior of each atom depends on the distribution of electrons among its levels. You have seen how a potential well confines the electrons within it and forces them to exist in one or other of a number of states, of amplitudes that provide a set of different permissible energy levels for the electrons within the atom.

"The number of electrons within any atom is decided by the nucleus. A carbon nucleus has a positive charge six times as great as the negative charge carried by each electron, and so it can attract six electrons before its charge is masked and neutralized. It is the nature of any system, whether a large scale *classical* system or a small scale *quantum* system like this atom, to end up with the lowest potential energy that it can and so, if there were no Pauli Principle, this would mean that all six electrons in the atom would be in the very lowest level. The Pauli Principle prevents this, for each electron must be in a different state, one that is in some way distinguishable from all others. In order to understand atoms you must know how many electrons may be held by each level."

The deep potential well disappeared, to be replaced by a large plywood cutout with a trumpet-shaped outline like a cross-sectional

---

[2]The Standard Model of elementary particles will be treated in some detail in *The Particle's Progress* by the present author. The style of the book will be broadly similar to this one.

view of the downward curving well. This supported a series of small shelves or ridges that corresponded in position to the energies of the various levels. There was one much lower than the rest, while significantly higher up the others came at ever closer intervals, one above the other. In each shelf there was a number of small depressions, each suitable for balancing a rounded object like an egg or a coconut. The whole affair was painted in gaudy colors and surrounded by a flashing halo of electric lights. To one side there was a sort of display panel with flickering red letters, rather like the display of a digital clock.

"There you see before you the various energy levels which are provided by the electric field of the atomic nucleus. Roll up, roll up! Just throw some electrons into the field and *make an atom!*" The Clown reached behind the display and produced a large, bulging canvas sack. From this he extracted what Scrooge fancied he could now identify as electrons. He passed some to Scrooge, who obediently threw them toward the target with its various levels. The Spirit gave a rapid but extensive commentary as Scrooge made a succession of casts.

"Electrons have an intrinsic spin, an internal rotation which is a part of them. They may rotate in either direction, giving what are termed *spin-up* and *spin-down* electrons. These two states are different and this means that in fact two electrons may be put into every state defined by the potential well—one with spin up and one with spin down—and the overall state of each electron will still be different as required by the Pauli Principle. At the lowest level there is but one state, and consequently we can put in it just two electrons of opposite spin." Both of Scrooge's first two throws struck the board at the top, and the electrons slid right down it to end up together on towest shelf. The display by the side of the board briefly showed $_1$H and then steadied to display the symbol $_2$He. "Now no further electrons can go into the lowest level, because this is now full. They must now go into the next higher level."

Scrooge continued to toss electrons toward the target. His third throw hit the top of the board as did the two before and began to slide lower, but this time it stopped at the second shelf from the bottom and lodged there. The display panel now announced that Scrooge had produced $_3$Li, an atom of the metal lithium.

"As the second level is quite a bit higher in energy, so this third electron has not fallen so far or lost so much potential energy as did the first two. It is much less tightly bound. At this energy level there happen to be a greater number of different states available. It is now possible for the electrons to have some orbital motion, some *angular momentum*, around the nucleus and there are states both with and without this orbital motion. They all give the same energy for the electron. You would have to perform a fairly complicated calculation to show what angular motion is possible at each level and as a consequence to find the number of states available, but the outcome is that eight electrons may sit at this level. These various levels are often called *shells*, because of course the atom is in reality three-dimensional and the electron states form shells that surround the nucleus."

Each subsequent electron which Scrooge tossed toward the board, from number three to number ten, slid down into this same level. He looked along the line of electrons that now rested on the second shelf and noticed that, although all the shelves had seemed completely flat and level to begin with, not all of the electrons on the 'level' were now at the same height. He commented on this discrepancy.

"You are quite right. The states within the potential well are defined solely by the electric charge on the nucleus and have the same energy on each principal level, quite independent of any angular momentum. As more and more electrons are added, the situation

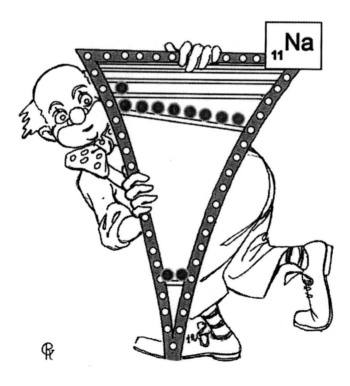

changes slightly. Each and every electron carries an electric charge, and so they interact with one another, as well as each interacting independently with the electric field from the nucleus. The effect produced by the electric charges on the other electrons is not as great as that of the nuclear charge, but nonetheless it is there and it does alter the energy levels in atoms that contain many electrons."

The eleventh electron which Scrooge pitched slid down the board only so far as the third level and there it stayed, in lonely isolation, while the display said that he had now created $_{11}$Na. "This brings us to Sodium, a chemical element whose atoms contain eleven electrons. Ten of them can fit into the first two levels, but the eleventh one is left on its own. It is the only electron to be in the third energy level, and so it is much less strongly bound than the others. The process of constructing larger and larger atoms continues in the same way. This third level also can hold eight electrons, so there will be another jump in energy for the last electron as you move from eighteen to nineteen electrons, from Argon to Potassium. I think the general trend should be clear by now."

Scrooge found that he had now thrown all of the electrons which he had been given, and the Clown tied up the neck of his sack and tucked it away behind the board.

"Atoms are of such a size that their nature is determined by quantum effects. Larger objects, the classical objects which you think of as normal, are built up by combining together many atoms. Everything you see around you in normal, everyday life is very large compared with the size of an atom, and there is an enormously greater variety and abundance in the substances you see around you than there are different types of atom. There are not, after all, very many distinct atoms, only about a hundred in all, and most of them are rather rare.

"In order that a limited set of small atoms may combine together to construct the world you see around you, *two* distinct ways of combining atoms are necessary. You need some way of putting atoms together in *very* large numbers to build up to solid objects large enough to be seen, because atoms are *very* small. You also must have some way of combining atoms together in a particularly intimate way so that they produce *molecules*, compound objects that serve instead of atoms to form the building blocks for different materials. Although there are relatively few types of atom, there are many, many different molecules. It is these molecules, rather than the individual atoms themselves, that are responsible for the properties of substances. This is chemistry.

"Take salt, for example." The Clown reached into his pocket and produced an enormous salt cellar. He upended it and looked in apparent surprise at the stream of salt that cascaded onto the ground. Reaching down he picked this up and threw it over his left shoulder. Behind him the white powder separated into a glittering metallic dust and a rather nasty looking yellow gas.

"Salt is sodium chloride. Sodium is a metal, chlorine a gas. Their individual atoms combine in a one-on-one basis to form something that is quite different from either. Why do they do this? What causes their atoms to stick together so intimately that it is the molecule, the combined object, that acts as the basic unit? Observe an atom of sodium and an atom of chlorine."

Cutouts representing both atoms could be seen side by side. One was labeled **Na**, the other **Cl**. The Clown told Scrooge that these were the *chemical symbols* for sodium and chlorine. The two boards looked very much the same, save that the shelves for the chlorine board were all at much lower energy levels than the sodium one. The Spirit told Scrooge that this was because the nuclear charge on the chlorine nucleus was greater, and it pulled the potential well down that much deeper. A chlorine atom has seventeen electrons, a sodium atom has only eleven, and their nuclear charges are in proportion.

"You see there that the sodium atom has one electron sitting exposed on its own, the only one in the third level; and consequently it has a much higher energy than the others in that atom. It must feel very isolated and exposed. Look now at the chlorine atom and see that here the third level is almost full, indeed only one electron is needed to complete the set. You see before you two atoms, one of which has an electron too many, one an electron too few. Would it not be a kindness for them to share? Would it not be better if the sodium atom were generously to donate its surplus electron to the chlorine, complete that atom's third shell, or energy level, and emptying its own? In the process it would reduce the overall energy of the system, because the electron would have a lower potential energy in the last available state of chlorine than in the first state of sodium. Both states are on the third level, but that in chlorine is more tightly bound because the greater nuclear charge makes the energy of all levels that much lower."

Glancing furtively round, the Clown sidled closer to the two atoms in a particularly unconvincing attempt at nonchalance. While he whistled innocently and gazed very obviously in the opposite direction, he reached out behind him and swiftly moved the isolated

⟪ ELECTRON ENERGY LEVELS AND CHEMISTRY ⟫

Chemical behavior is produced by the electrons in atoms. The nature of an element depends entirely on the levels occupied by the electrons, especially the highest occupied level, known as the *valence level.*

The whole vast range of chemical behavior depends on the existence of the different energy levels which are available for electrons within an atom and the operation of the Pauli Principle. The Pauli Principle limits the number of electrons which may fit into each level, and the atoms "fill up from the bottom." Without the Pauli effect all electrons would be in the lowest level and all atoms much the same.

Energy levels in sodium and chlorine

An example of chemical bonding is given by the ionic bond in sodium chloride: **salt.** Sodium has one isolated electron in its highest level, and chlorine has one space available. Because of the higher nuclear electrical charge in chlorine the electron binding is greater and the levels lower than in chlorine. Energy is released when the last isolated sodium electron moves over to fill the space in chlorine. Consequently this is what happens, producing *ions*, atoms with the "wrong" number of electrons.

The result is that the chlorine has an electron too many and is thus negatively charged, while the sodium has an electron too few and is positive. Electrical forces then hold together the ions of opposite charge.

electron from the upper shelf in the sodium display to the single remaining space in the chlorine.

"Now isn't that better? The total potential energy has been reduced, and that is always the most favored result. What you see here is what happens in practice to real sodium and chlorine atoms. The amplitude for the electrons explores every possibility and will settle on the configuration that gives the lowest potential energy. Now that an electron has been moved from the sodium atom to the chlorine atom there is an obvious consequence. The sodium atom has now only ten electrons but a positive charge on its nucleus which is equivalent to eleven electrons. The chlorine atom now holds eighteen electrons but has a nuclear charge equivalent to only seventeen. Both atoms have become ions. They are no longer electrically neutral atoms but are, respectively, a sodium ion with a positive electrical charge and a chlorine ion with a negative charge. Because the charges are opposite they are attracted to one another and so bound together to form one composite object. This is a sodium chloride molecule, the substance you know as salt. It is held together by the electrical attraction between the two electrically charged ions, by an *ionic bond*.

"There are many ways that atoms can combine into molecules. All depend on the behavior of *valence electrons*, which are those in the outermost shell, the one that is not completely filled. The electrons in the inner shells have the best deal that they could get and are not tempted to change. Some molecules are very simple, like the salt molecule, which consists of but two atoms. Some are extraordinarily large and complex, containing hundreds of atoms. Such are *organic molecules*, to be found particularly in living organisms. Sodium and chlorine combine readily because they are at the extremes of chemical activity. One has a single surplus electron, and one lacks a single electron needed to fill a shell. They clearly benefit by passing an electron from one to another. Organic molecules are based on the properties of the carbon atom, which is right in the middle of a shell. Carbon has four valence electrons. Four more than an empty shell, four less than would be needed to fill the shell completely. It is not clear whether it is to the advantage of a carbon atom to lose or to gain electrons, so what it does is to *share* them with another atom.

"One of the simplest carbon compounds is methane, in which one carbon atom combines with four hydrogen atoms. It is not obvious in this case which way the electrons should transfer. If each hydrogen atom passed its single electron to the carbon, then the carbon shell would be filled by eight electrons and the hydrogen atoms would have no isolated electrons—indeed they would have no electrons at all. If the carbon atom distributed its four electrons among the four hydrogen atoms, each hydrogen atom would now have two electrons, enough to fill the first shell completely and the carbon would have no electrons left in its outermost shell. Which is the best choice? Which option does the carbon atom choose? If the energy of one set of levels was much lower than another then the electrons would move into that one, but with similar energies the choice is far from clear. This is a quantum system, so it makes no choice; or rather, it chooses *both*! The electrons are in the hydrogen atoms and *also* in the carbon atom." If an electron can go through two holes in a screen it sees no problem in being in two atoms at the same time, and this is what it does. Such is a *covalent* chemical bond, and it is able to build up complex molecules that contain many atoms of the same type.

"Even the largest molecules, which may contain hundreds and hundreds of atoms, are still pretty small. They are all far too small for you to notice one of them. The things you work with, large things like pinheads that you would actually be able to see, contain a truly enormous number of atoms and molecules. The number is so large that most materials appear to be smoothly continuous. However things may seem, no substance is completely uniform on all scales. Magnify a portion of it sufficiently, and you would find that any solid object is made from huge numbers of molecules stuck to one another. Why do they stick together? It is again because of electrical attraction.

"The reason that there should be an attraction is not obvious because atoms are electrically neutral. The positive charge on the nucleus is exactly canceled by the negative charges on the atomic electrons, and the total charge is precisely zero. However, although the negative charge carried by the electrons may be equal *in magnitude* to the positive charge carried by the nucleus, the two are not *in the same place*. The positive charge is concentrated in the tiny nucleus, the negative charge spread over an electron cloud. Electrical forces depend not only on the size of the electrical charges concerned but also on how far apart they are. The farther apart, the weaker the force.

"If two similar atoms, or two molecules for that matter, come close to one another, then the presence of the other may distort slightly the shape of the electron cloud. It may be that in consequence the repulsion between the similar charges in the two nuclei and the negative charges in the two electron clouds will be less than the attraction between each nucleus and the electron cloud in the other atom. If it is *possible*, then this is what *will* happen. Attractive forces lower the potential energy, and systems invariably choose the state with the lowest potential energy. There is a small imbalance between the repulsive and the attractive forces. It is small compared to the attraction between electrons and nucleus in an atom. It is small compared to the attraction between ions in a compound, and it is only perceptible when the atoms are very close together and the different positions of the charges within an atom produce a noticeable effect. The force produced is relatively insignificant, but it suffices. If atoms are sufficiently close they are attracted to one another and stick together."

Scrooge found himself enveloped in a cloud of atoms. He was seeing them now in normal space, and as before they appeared as fuzzy spheres that drifted around him. Two collided and remained in contact. More joined to these, and in next to no time there was a great clump of these balls stuck together, like the vision of a solid that he had earlier been shown by the Shadow. As in that case the atoms all fitted closely together; and like eggs in a packing case, they arranged themselves into a regular array so that they could crowd together the more closely. Larger and larger grew the conglomeration of atoms, and Scrooge's viewpoint withdrew to encompass this greater expanse. As a consequence he could see less detail and the individual atoms were less evident, but he could see clearly the lines and planes of atoms as they extended in a regular *crystal structure*. As more and more atoms stuck to the developing body and his view expanded with it to include it all, he found he could no longer see the individual atoms. The body now seemed homogeneous and continuous. Its discrete structure was no longer visible, and the atoms which still condensed upon it now had the aspect of a continuous vapor.

More and more remote grew Scrooge's viewpoint, and the atoms within the body before him fell ever farther below the threshold of his perception. At last the accumulation of atoms ceased, and he was looking from a little distance at the hard and shiny surface of a solid body, one which he recognized as a polished metal paperweight that usually sat upon his own desk. It was still upon his desk, and about him were all the familiar contents of his office. He was in his own place of work. Atoms or no atoms, it looked just the same as ever. He looked around with satisfaction and remarked to the Clown who had accompanied him, "Perhaps atoms may indeed experience all these strange quantum effects of which you have spoken, but on the familiar scale of life they seem to have no effect."

"Wrong! That is far from the case," replied the Spirit. "The properties of the atoms govern the world that you see about you. Those properties, and in particular the chemical behavior of atoms, is determined by quantum effects and the everyday objects about you are all built from such atoms and share their chemical properties.

"Admittedly, most quantum behavior is clearly evident only on a very small scale but not invariably so. There are quantum effects that can and do operate on a large scale. This may be shown in the reduction of amplitudes—that extraordinary process whereby an observation determines the true system present. This is among the strangest of quantum effects and can operate over considerable distances.

"The simplest demonstration involves the properties of spin, the strange internal rotation possessed by many particles. As I have already mentioned, the spin of electrons is tightly restricted. If you measure the spin along any direction it may point one way or another, with no half-way values. Classically, you might expect that this spin could point in any direction, and that its component, the portion which would lie along any axis you cared to choose, could take any value. In a quantum system this is not so. Only two choices are available. The spin component either points in the direction you chose or it points the opposite way. There are no other choices."

The Clown held up a large colored ball, a large-scale cartoon representation of those fuzzy particles which Scrooge had seen so recently in the microscopic world. Abruptly this picture became blurred and, like a fade between two film sequences, was replaced by two balls of a different color that moved steadily away from one another. They moved slowly so that Scrooge could see the arrows painted on their sides and had time to note that the arrows on the two balls were pointing in opposite directions.

"Sometimes particles will decay to eject two new particles," explained the Spirit. "These travel away in opposite directions because they must conserve momentum. If the original particle has no spin, then of necessity the total component of the spin must be zero. The two particles produced have this same total component; so if they do each possess spin, then they must have components in opposite directions, so that these cancel, and the *total* is still zero. So far there is nothing particularly strange or paradoxical. The interesting aspect arises when you consider that either of these particles might have its component of spin in either direction. They must be opposite to one another, but each could point either way. As they *could* be in either direction, of course they will have a sum of amplitudes for the two directions. That is standard for quantum systems."

Now Scrooge could see that each of the cartoon-like balls had two arrows displayed upon its side, like two images projected upon the same screen. As a consequence the two now looked completely alike.

"Even so, all is well until the direction of the spin is *observed* for one of the particles. In that instant the direction of the particle's spin has become definite, and the sum of two amplitudes has reduced to one. This is strange, but no more strange than we have described it on previous occasions. What is new here is that we know the spins of the two particles *have to be* opposite, and so when the superposition of wave functions reduces for one of them it must automatically reduce for the other also."

The Clown stretched out his arms to grasp each of the two particles, which had now separated by an arm's length. When he opened his hands each particle could be seen to bear but a single arrow, pointing in opposite directions to one another.

◖ EPR Paradox ◗

This Einstein Podolsky Rosen paradox is a paradox in that things seem to be happening that require Nature to communicate faster than the speed of light. The situation is roughly as described in the text. When some particles decay they produce two particles that have spin and move off in opposite directions. The particles must have spins pointing in opposite directions, but each might have either direction of spin. Quantum mechanics says the spin direction for a particle becomes definite only when a measurement is made. When this happens it fixes the spins of both particles even if the other is a long way away. Measurements of both spin directions would show they were correlated.

It would seem to make sense if somehow the particles could decide which would point in which direction when they were produced and then simply carry this information with them. Should the directions along which the spins are measured be decided independently for the two particles, *Bell's theorem* has shown that there is a limit to the correlation that can be produced by any information carried with the particles. Experiments support quantum mechanics and show greater correlation than would be allowed.

The bottom line is that the particles act together more completely than they could by carrying information with them or exchanging it at the speed of light. There appears to be some sort of bookkeeping for the quantum amplitudes that takes place faster than the speed of light. Strange, but apparently true.

"What is really striking is that this reduction for both particles *must* happen, *no matter how far away the particles may be!*" The Spirit's voice rose to a dramatic climax but also sounded strangely remote. Scrooge looked around, but saw no sign of his erstwhile constant companion. Apart from himself the room appeared to be quite empty. Had the Spirit shrunk once again into the microscopic world of the atom? He glanced casually out of the window behind his desk, then abruptly looked again. Outside it was dark, and the heavens were full of stars. Against this vast astronomical backdrop he saw the diffuse outline of his companion spread across the night sky. His arms were stretched wide, and his two hands were clenched upon distant galaxies, inconceivably remote from one another but still connected in some incredible fashion by the Long Arm of the Clown.

As Scrooge looked on in wonder at this celestial manifestation, the vision of the Clown faded from view, and he became aware that his companion was again in the room with him and again lecturing him. "What is really remarkable about this long-distance reduction

of the wave functions," pursued the Clown, "is that it can happen in two different places for events with a *space-like* separation! The correlation is better than could be achieved by sending any sort of message between the two particles at the speed of light! There is some nonlocal effect which would appear to be in contradiction to the theory of relativity. In fact the contradiction is not as absolute as it might seem. The theory of relativity maintains that no message and no energy may travel from one position to another at a speed greater than that of light. Neither happens here. No energy is transmitted, so there is no violation there. If you look closely at the circumstances, you may see that no message can be transmitted either. It might appear that here at last is the way of sending messages faster than the speed of light that has been so long sought by writers of science fiction. Previous possibilities that have been mentioned were inadequate because they would not actually travel faster than light. In this case it does appear that light-speed is not a limitation, and the amplitude can and does connect events separated by space-like intervals. The problem this time is that you cannot send messages.

"You might think that you have a way of sending a sort of Morse code. You have two directions for the spin, which you may interpret as the dot and dash of the code. Depending on which spin you choose, you know that the opposite becomes reality for the distant particles. Apart from some tricky engineering problems, the difficulty is that you cannot choose which spin you will observe— whether you will have a 'dot' or a 'dash.' Which possible state of the particle is observed is *purely random*. There is no way in which you can *make a choice* and consequently no way in which you can *impose any message* upon the reduction of the amplitudes. The existence of correlations over space-like intervals is well established."

Scrooge was impressed by the scope of this effect. "Truly that is a demonstration of quantum effects on a large scale," he admitted. "The thought that the amplitude for those two particles may extend over the distance between stars and that they are still in some sense part of the same whole is truly staggering. This must assuredly be the grandest vision of a quantum amplitude which you can devise.

"Well, actually, NO," replied the Clown. "That picture of two particles still bound together by some tenuous thread of quantum amplitude over astronomical distances pales totally into mundane insignificance beside the overwhelming concept of the 'Many Worlds' theory. This is the final, logical extension of the idea of superposition of amplitudes. We know that an electron can be in a superposition of amplitudes, for we can see the interference between the different amplitudes that results. All electrons behave in this way, and we see that a group of electrons or other particles will exist also in such a superposition. It seems to be an absolute general rule for all quantum systems that whenever any result or action is possible there is an amplitude present for that action *as well as all others.*

"This superposition reduces to one unique instance whenever an observation is made, but what we can make such an observation? If the world is a structure built of many particles and all the particles follow the rules of quantum physics, then surely a large number of them, such as the world, must follow these rules also. Any observer is part of the world, so you might argue that he or she should show the same quantum behavior. Whenever an observer looks at a superposition of wave functions, then there are many results which *might* be observed. In such a case a quantum system will not select one of them, but will be in a superposition of states, each of which corresponds to seeing one particular result. If the whole world is quantum, then this argument may be continued indefinitely, and nothing ever uniquely *happens* or does *not* happen. In this picture the world is best described by a vast amplitude in which *nothing* is excluded. It is a gigantic superposition of everything that could ever possibly have happened. Nothing, no possibility, however remote, would ever have been excluded. It is often said that 'anything can happen.' In this view there is nothing that cannot, *has* not, happened in the sense of being a legitimate part of the grand universal amplitude.

"You might argue that this obviously cannot be because you are aware of only one single result in every case. Consider, however, that in this picture the amplitude contains a superposition of states for everything in the world, and this includes *you.* There will be an amplitude for many different versions of you, each of which corresponds to one particular amplitude for each effect, and so will see unique results from observations because those observations are just the amplitude which goes with *your* amplitude. Each of these versions of you would be unaware of the others. Such complex structures do not show evident signs of interference, and so each would feel itself to be unique. The Many Worlds picture is actually quite economical in its assumptions, though at the expense of being rather extravagant with universes."

Somehow the Spirit caused the distinction between these amplitudes to blur, so that Scrooge was vaguely aware of many different versions of himself. Not of himself only, indeed, as each amplitude carried all the people and things that were in any way affected by

◖ Many Worlds Interpretation ◗

There is an inherent difficulty in the interpretation of quantum mechanics.[3] A central feature is the idea that anything possible is included in the overall superposition of amplitudes, which describes the system. It is difficult to avoid this idea because it allows the existence of interference between amplitudes, and such interference is clearly seen. On the other hand we know that when we look at something we see only *one* behavior in any case. This leads to the reduction of wave functions by an observation that gives only *one* result.

The difficulty is that quantum systems do not normally give only one result, and this suggests that whatever is making the measurement is not a quantum system. Where is the boundary line at which quantum behavior stops? The Many Worlds picture says that no such boundary exists and that *everything* is described by a superposition of quantum amplitudes. This amplitude includes all possible observers. In this theory there are as many versions of each observer as there are possible results for every observation that has ever been made. This concept gives a fairly self-consistent interpretation of the so-called measurement problem, though it does seem rather extravagant.

Scrooge's existence, and so it was different versions of the whole Universe that he beheld.

The number of these possible worlds was truly, inconceivably, vast. Most of them differed but little from one another. An atom in one might move in a slightly different direction from that atom in another. The number that differed no more than this was already gigantic, since there are many atoms in the world, let alone in the Universe. Greater by far was the number where the behavior of two atoms differed, or three or four. Great as was the number of universes that differed to such insignificant degrees, there were others that had followed increasingly differing paths. There were huge numbers of these, distinguished only by insignificant differences; but among the whole incredible range were some that showed serious and surprising differences. The whole immense range of what *might*

[3]The "Many Worlds" picture is just one of the different suggestions which have been made for interpreting this "measurement problem." This suggestion and others are treated at more length in the author's earlier book *Alice in Quantumland*.

happen, however unlikely, was included somewhere among the almost infinite extent of the many worlds. Some instances had a certain familiarity.

* * * * * * * * * * * *

* *

Scrooge pushed back his chair and stood in his place at the center of the top table, basking in a glow of self-satisfaction as he prepared to address The Association of Financiers.

* *

Scrooge's presence upon the scene was far from evident as nothing of him remained but one overgrown tombstone in a forgotten corner of the graveyard, bearing upon it the chiseled name SCROOGE.

* *

* * * * * * * * * * * *

Some amplitudes were really improbable, but any possibility, however unlikely, contributed with some tiny probability to the whole great superposition of the many universes. There was nothing that *could* happen, however implausible it might be, but that some amplitude existed in which it *did*.

* * * * * * * * * * * *

* *

To the tumultuous delight of the great crowd of screaming fans, Scrooge strutted out upon the brightly lit stage in the center of the great crowded arena. All around him he heard the hysterical cries of 'SCROOGE,' 'SCROOGE' as he continued the triumphal tour of his chart-topping group *"Dickens' Grandchildren."*

* *

Fully aware that the whole world was watching him through the eyes of the television camera, Scrooge took the last step from the ladder onto the surface of the Moon, and the television channels carried worldwide his immortal words: "One small step for man, but a heck of a long way stuck inside this damned spacesuit."

* *

* * * * * * * * * * * *

In some cases the amplitudes in the different universes were effectively indistinguishable, and there was interference between them. In most cases they were distinct and distinguishable, though the differences were generally very small. In such cases the universes had no effect upon one another. No one of them had any knowledge or awareness of the others' existence. For any Scrooge selected from the almost infinite total, it was as if the others did not exist. As far as each of them was concerned, that particular Scrooge was the *only* Scrooge, a truly unique individual. For us, viewing in this narrative the whole superposition of states that composes the universe, indeed the universe of universes, there is no reason to say that any one or other of these amplitudes represents the true Scrooge, since all are equal members of the grand superposition. For practical purposes, for the purpose of continuing the story, we must make some choice and so we choose. . . .

****

Scrooge awoke. He saw at once that he was in his own bedroom and that the early sunlight was streaming through the cracks in his window blind. It was morning. The long and strangely eventful night was finally over.

Not the Man He Was

Scrooge woke up with a start. It was morning, and the sunlight was streaming into his bedroom through every gap in his window blind. The night was over at last, and all three of the promised Spirits had visited him. Although in their company he had traveled toward the limits of space and time, the whole experience had been contained for him in a single night. Now the night was over, and everything was back to normal. Or was it? Not entirely, for Scrooge was not quite the man he was.

Certainly he rose early, as was his wont, made a quick breakfast, and was out of his apartment and on the way to his office with little time wasted. Certainly he intended to use the day as always to

his best financial advantage. Despite this, however, Scrooge was not the man he was. As he made his way to his office he looked around with new eyes. He looked up at the overcast sky (this too had returned to normal) and was aware that, above the clouds, planets, stars, and galaxies were all rushing in their various ways, and that the seemingly stationary earth beneath his feet was no more at rest than any of them. He looked at the early beams of the Sun that penetrated the clouds and was aware of the energy which the Sun fed to the earth each day. This gave the energy needed to drive the whole great system of the World until eventually that energy became locked in the great background of unavailable heat so that fresh energy was required. He looked at the windows of the shops and the steel gateway to the underground station and was aware how everything he saw was composed of atoms and of the strange quantum world which those atoms inhabited.

As he looked at everything around him Scrooge seemed to hear behind it the pipes of Pan playing complex tunes of chaos and of developing structure, with superimposed on it all as a universal background the strange all-pervasive notes of the quantum amplitude.

On his way to his office Scrooge chanced to encounter his cousin, the same cousin who had had such an unsatisfactory meeting with him but the day before. To this young man's vast surprise, Scrooge accosted him and thrust into his hand a considerable donation for the cause he had previously so firmly rejected. For Scrooge, though no less a man of business, had discovered a whole new reality in the real world and had developed a great interest in discovering more of its nature. As time went by, he became known for it among his circle of acquaintances and indeed, it must be admitted, was sometimes felt to be a bit of a bore on the subject at cocktail parties.

And, as he had suggested to Marley's ghost, he *did* take out a subscription to *Scientific American*.

Index

Printed in the United States
88638LV00002B/76-78/A